Innovation in
African Agriculture

Innovation in
African Agriculture

Arthur J. Dommen

Westview Press
BOULDER & LONDON

Cover sketch by Carl Mabbs-Zeno.

Westview Special Studies in Agriculture Science and Policy

Published in 1988 in the United States of America by Westview Press, Inc., 5500 Central Avenue, Boulder, Colorado 80301

Library of Congress Catalog Card Number: 88-14365
ISBN: 0-8133-7648-3

Printed and bound in the United States of America

The paper used in this publication meets the requirements of the American National Standard for Permanence of Paper for Printed Library Materials Z39.48-1984

6   5   4   3   2   1

Contents

Tables

viii

Figures

x

Acknowledgments

The author is an agricultural economist with the Economic Research Service of the U.S. Department of Agriculture. The research reported in this book was undertaken in line with the author's responsibilities as Research Section Leader, Africa and Middle East Branch, International Economics Division, from April 1980 until the reorganization of ERS in July 1987. The background in the economics of African agriculture acquired in the course of these years of reading, discussing, and seeing on the ground was invaluable. Needless to say, the author assumes sole responsibility for the contents of this book. It is hoped the book will furnish those who persevere in an area of investigation where fewer contemporaries write with authority than is often assumed to be the case with some useful insights and a provocative hypothesis. It is not the author's intention to implicate the Department in any way with such errors or omissions of fact or analysis as the book may contain.

Several persons took the time to read different drafts of the manuscript and a short paper presenting its hypothesized production function at the 1987 annual meeting of the Eastern Economic Association. They all deserve thanks. The author wishes to express his appreciation particularly to Professor Ian Hardie of the University of Maryland, who at a crucial stage of conceptualizing the problem suggested a functional form treating conservation of equilibrium biomass as an output rather than an input, allowing the analysis to move ahead rapidly; and to Dr. Balu Bumb, now of the International Fertilizer Development Center, Muscle Shoals, Alabama, who kept him straight on mathematics and provided needed encouragement throughout.

Arthur J. Dommen

"Fertility may be the result of the use of intensive methods of land utilization and not *vice versa*."

—Ester Boserup
The Conditions of Agricultural Growth, p. 19

Introduction

Africa is the continent where the earliest remains of the human race have been found. Africa's agriculture is very old, dating back at least 5,000 years in the case of yam cultivation in the "Yam Belt" of Nigeria and Cameroon. The Ethiopian highlands and the rivers of West Africa have been identified as centers of domestication and diffusion of important crops grown today.

Yet African agriculture appears to be singularly resistant to change. Its crops and tools are largely the same as those used by farmers thousands of years ago. Techniques and practices of tilling the soil encountered in African villages date back many generations. Attempts from outside to impose change at a more rapid pace have invariably deceived their proponents. The evaporated dream of the French colonial government in the 1930's to bring one million hectares under rice and cotton cultivation in the inland delta of the Niger River and the British-designed "groundnut scheme" in East Africa after World War II are well-known examples of failure.

More recently, declining per capita food production in Sub-Saharan Africa over the past 25 years has led to a debate about whether agricultural research has failed to produce new plant and mechanical technologies appropriate to African conditions, or whether such technologies exist but have simply not been adopted by farmers for a variety of reasons having to do with inappropriate policies, lack of incentives, rural-urban migration, and absence of input delivery systems and extension services.

This debate, with its implications for the directions of future research, for the possibilities of technology transfer, and more generally for the prospects of revers-

ing present aggregate production trends in Sub-Saharan
Africa, appears to have reached a deadlock. Agronomists
listening to economists feel the latter minimize the
extent of the constraints, especially paying mere lip
service to the need for resource conservation, and too
readily assume that once economic conditions are right,
African agricultural production will "take off." Economists in their writings frequently convey the impression,
intentionally or unintentionally, that plant scientists
are to blame for the failure to develop appropriate
technology for African agriculture. African farmers'
views in all this, unfortunately, are seldom considered.

An observation about low-resource agriculture in
Africa frequently made by economists is that factor (or
input) productivities are low. If this is true of all
factors, then the scope for factor substitution is very
limited, and technological change raising the productivities of land, labor, and other factors would seem to be a
necessary condition for achieving sustained increases in
food production. Indeed, a recent compilation on African
agriculture takes this as one of its two basic premises.[1]

This book takes a fresh look at the input-output
relations of what will be called here low-resource agriculture in a way that moves the debate off dead center.
Low-resource agriculture is defined as agriculture that
relies primarily on internal inputs (farm-produced seed,
family labor, simple hand tools, manure and organic
wastes, minerals from ash) rather than on external inputs
(commercially-produced seed, hired labor, power tools,
chemical fertilizers).

Low-resource agricultural systems may be said to be
based on the following features of African rural communities: (1) an ability to work with the environment rather
than attempting to override it; (2) a deliberate seeking
of diversity of micro-environments; (3) the purposeful
selection throughout the production period of crops
planted and cultivating practices used and the integration
of livestock into the system as a means of maintaining
soil fertility; and (4) the deliberate staggering of
outputs in space and time.[2]

In the view of many who have studied them scientifically, such systems are very flexible and possess great
capacity for self-adjustment at the initiative of the
community. They do not take kindly, on the other hand, to
change imposed from without. In fact, imposition of outside change frequently upsets their capacity for working

with the environment, and may lead to the destruction of the agricultural production system itself.[3]

This description intentionally places low-resource agriculture in Sub-Saharan Africa in quite sharp contrast to commercialized agriculture, which depends on a controlled (and, if possible, uniform) environment and standardized packages of recommended inputs, and is based on monoculture. It should be noted that the so-called Green Revolution in Asia and Latin America in the 1960s and 1970s was much closer to the latter than to the former in terms of these features.[4]

Of the two problems of agricultural production in Africa--the "transformation" problem of converting production to large-scale commercialized production where this is possible on the pattern of irrigated cotton production in the Gezira or of mechanized wheat farming in Zimbabwe, and the "improvement" problem of assisting millions of African low-resource farmers to raise their output marginally without a major transformation of the environment--this book is concerned only with the latter. There are those who assume, implicitly or explicitly, that Africa's agricultural production problem can be solved by "transformation" alone; but I am not willing to make such a broad mental leap.

NOTES

1. "The second premise of this book is that technological change in agriculture is a necessary condition for achieving sustained increases in food production." John W. Mellor, Christopher L. Delgado, and Malcolm J. Blackie (eds.), Accelerating Food Production in Sub-Saharan Africa (Baltimore: Johns Hopkins University Press, 1987), Chapter 1, p. 5.

2. There now exists a fairly extensive body of documentation on low-resource agriculture in Sub-Saharan Africa, to which agriculturalists, economists, geographers, anthropologists, and other scientists from numerous countries have contributed. This literature is not always easy to find, and language may be an obstacle, but an idea of its richness can be had by examining, for one region alone, the mimeographed reports of the workshops on Sahelian agriculture held at Purdue University in 1979 and 1980; the collected papers presented at the

workshop in Ouagadougou, April 2-5, 1985, edited by
Herbert W. Ohm and Joseph G. Nagy, Appropriate
Technologies for Farmers in Semi-Arid West Africa (West
Lafayette, Indiana: Purdue University, 1985); and David
W. Norman, Mark D. Newman, and Ismael Ouedraogo, Farm and
Village Production Systems in the Semi-Arid Tropics of
West Africa (Patancheru, India: ICRISAT, October 1981).
This large body of literature remains, for the most part,
undigested.

3. This vulnerability of Sub-Saharan African
agricultural systems to destruction from outside,
particularly to introduced techniques of production that
destroy soil structure, has resulted in the
mischaracterization of African agriculture as "fragile,"
whereas, in fact, its most remarkable characteristic is
the flexibility with which it uses internally available
resources.

4. Which is not to say that the Green Revolution was
not largely on a smallholder basis, like African
low-resource agriculture.

Efficiency

Introduction to Part One

Since the publication in 1964 of Theodore W. Schultz's <u>Transforming Traditional Agriculture</u>, economists have generally accepted the thesis that farmers engaged in traditional agriculture are producing at high levels of efficiency, measured by marginal returns to resources in alternative uses. An important corollary of this thesis has been that significant increases in output cannot be obtained by reallocating existing resources, but only through technological change that fundamentally restructures the productivities of these resources. The success of the Green Revolution in Asia and Latin America in diffusing high-yielding varieties of wheat and rice developed by international research centers seemed to offer convincing empirical proof of the correctness of the Schultz thesis.

Nevertheless, Sub-Saharan Africa has so far not followed this technology-based path to agricultural development. African farmers, it seems, have resisted adopting available new biological and mechanical technologies even though such technologies have demonstrated their "superiority." Is Africa's systematic failure to generate significant positive growth rates of agricultural, and in particular food, production per capita attributable to some flaw in these technologies, or to peculiarities in the way resources are used in African agriculture?

At first sight, African farmers appear to be anything but efficient. The following description by De Schlippe sums up the impression made on many Westerners by an African farm even today:

When one enters a Zande homestead for the first time, the impression is that of complete chaos. The courtyard is shapeless or roughly circular or oval. The huts in it are scattered. Crops, food and household belongings may lie about the courtyard, or be piled on to the veranda of a hut in what seems to be a most disorderly fashion. Worst of all, no fields can be seen. The thickets of plants surrounding the homestead seem as patchy and purposeless as any wild vegetation. It is impossible to distinguish a crop from a weed. It seems altogether incredible that a human intelligence should be responsible for this tangle.[1]

But this impression is misleading, as will be shown. Most investigations of the efficiency of African farming have focused on the effect of technological innovation on farming patterns. Matlon and Newman, for instance, were concerned in their study of northern Nigerian farmers with the effect of income distribution on small farmers' abilities to adopt new technologies.[2] Such investigations have often thrown light on the sources of African farmers' efficiency in traditional production, so that it may be said with confidence that the Schultz thesis applies to African farmers. This book will be concerned with the effect of these sources of efficiency on technological innovation and its possibility, rather than the reverse.

NOTES

1. Pierre de Schlippe, Shifting Cultivation in Africa: The Zande System of Agriculture (London: Routledge & Kegan Paul, 1956), p. 101.
2. Peter Matlon and Mark Newman, "Production Efficiency and the Distribution of Income Among Traditional African Farmers," African Rural Economy Working Paper, Michigan State University, mimeo, 1979.

1

Environment and Resources

In discussing low-resource agriculture in tropical Africa, it is necessary to relate the discussion to the physical environments which set the outer limits on how farmers organize resources for production. Two subcontinental agro-ecological zones will be distinguished in this book: the semi-arid tropics (SAT) zone and the tropical rain forest (TRF) zone. This is obviously too broad a scheme to satisfy many physical scientists. With the aim in mind of showing how African low-resource farmers behave in environments that are much too diverse to be captured by even the most detailed scheme, it will serve.

AGRO-ECOLOGICAL ZONES

The TRF zone is characterized by high annual rainfall and rainfall distributed throughout the year. The natural vegetation is dense forest. This produces a growing environment in which nutrients are recycled rapidly, being released from decaying vegetable and animal matter by the activities of decomposer organisms and almost immediately being taken up by the roots of trees and other plants. Thus, while the tropical rain forest represents an extremely efficient system, the stock of nutrients remaining in the soil is not large. A main characteristic of the tropical rain forest is its richness of species, combined with a low population density of any single species. Agricultural cash crops are mainly tree crops like oil palm, cocoa, and coffee which make relatively small demands on the soil because the amount of nutrients removed from the ecosystem in harvesting is not large.[1]

The SAT zone is characterized by seasonal amount and distribution of rainfall resulting in water regimes with marked rainy and dry seasons. These are of varied duration and intensity. The natural vegetation is grassland, with varying densities of scattered shrubs or trees, the ecological whole being described as savanna. The total content of available plant nutrients and bases in soils and ecosystems is small because soluble products of weathering and mineralization are readily and rapidly removed by leaching, soil structure is hardened by sunlight, and because nitrogen, sulfur, and other elements may be lost in fires. The basic food crops of the SAT are cereals, mainly sorghum, millet, and upland rice, supplemented by cowpeas, groundnuts, and other leguminous crops. These give way to starchy tubers, mainly yams and cassava, as the principal staple energy foods as one moves towards the TRF zone. Cotton is a major export crop of the SAT.[2]

Favorable Factors

Physical resources for agricultural production are not lacking in Africa. There is an abundant supply of land. Insolation in the SAT makes this zone potentially more productive of vegetable matter than most of Europe. Potential production of biomass at Samaru, northern Nigeria, is 2.8 times that at Rothamstead, England.[3] Abundant rainfall in the TRF zone is favorable for plant growth. On the other hand, high insolation imposes high evaporation demands on plants and depletes soils of moisture rapidly. Rainfall in the SAT is variable. Most rain comes from convective storms, producing high soil erosion rates. But the SAT of Africa, while risky for agricultural production, are no more so than the SAT of the other continents like northeast Brazil where viable agricultural economies exist.

Soils are known to be extremely variable in Africa. On the whole, they are not exceptional for their fertility, but neither are they infertile. Cultivation takes place on the desert edge, where water is the limiting factor rather than soil fertility, and the problems of the soils of the TRF when put under cultivation are well known. Soil fertility is known to vary widely within short distances.

CROPS

Many African societies are keenly aware of their agricultural heritage. Sorghum, pearl millet, and yams were all domesticated in Africa. The Ethiopian highlands, like the Zagros Mountains in the Middle East, were a region of rather abrupt altitudinal changes in vegetation zones that apparently proved favorable for experimentation in the domestication of crops and livestock. Teff and ensete are cereals that were domesticated there and are still regionally important as food sources. There undoubtedly existed other similar experimentation centers.

The varieties of staple crops planted in Sub-Saharan Africa are overwhelmingly those that have been planted by farmers for generations and are commonly called tradition-al varieties. These varieties stem from centuries of adaptation and transformation from earlier ancestors.[4]

Over generations, farmers have selected the varieties that are best adapted to rainfall and temperature regimes, soil moisture availability, length of growing season, photoperiod (number of daylight hours), and other influences. In Africa, this process of selection has been shown to extend to individual farmers, who select for local soil conditions and microclimate. For instance, women farmers in Machakos and Kitui districts in Kenya select pigeonpea seed of their own preference.

African crops extract scarce nitrogen and other nutrients from the soil with great efficiency. They often develop extensive root systems, drawing on a large area or depth of soil. Their vigorous growth suppresses weeds that compete for available nutrients.

Moreover, the traditional varieties of crops generally possess a high level of genetic variability, or heterozygosis. This is valuable to the farmer since it carries a measure of insurance against the risk of environmental stresses, insects, and diseases. If a crop becomes infected with a particular disease, for example, some of the strains in the crop population are likely to be susceptible, but others will be resistant, thereby limiting the extent of the crop failure.[5]

The main failing of traditional crop varieties is that their yields cannot be increased significantly. Spacing between plants cannot be reduced so as to increase plant populations per unit of land because of their large vegetative growth both above and below ground and because of their strong competition for nutrients and light.

Moreover, denser stands of plants would rapidly deplete the soil's nutrient content. Adding nutrients to the soil in the form of fertilizer may not result in higher yields because most of the newly available nutrients are not employed for increasing the edible portion (grain, tuber, and so forth), but for vegetative growth. Leaf area increases greatly and shading takes place, reducing photosynthetic efficiency. Height may increase unduly, causing lodging or stem breakage, and actually reducing yields.[6]

Although instances of importation of "exotic" crops are not unheard of in Sub-Saharan Africa,[7] there has been little tendency in modern times on the part of farmers to replace traditional varieties with higher-yielding varieties, as occurred in the case of wheat and rice in the Green Revolution. It is commonly stated that in the sorghum, millet, and upland rice area of West Africa probably less than 2 percent of the total area is sown to cultivars developed through modern genetic research.[8] The proportion must be roughly similar in other regions.

The record of crop technology transfer to Sub-Saharan Africa from other continents in modern times has been one of failure. With the possible exceptions of irrigated rice and a recently developed sorghum variety in Sudan, traditional African crop varieties have proved themselves superior to imported varieties. After 10 years of trials in which over 2,000 varieties were imported for trials in the mangrove swamps of West Africa, the West Africa Rice Development Association (WARDA) found only two varieties of rice that perform as well as the best varieties traditionally grown there.[9] Among some 7,000 sorghum introductions screened by ICRISAT in Burkina Faso, nine cultivars were found to be sufficiently promising in on-station trials to warrant on-farm tests. Of these, only two were found to be generally superior under farm conditions. Among some 3,000 millet entries screened, no superior cultivars have been identified so far.[10]

Principal Crops

Sorghum. Sorghum is one of the most widely diffused crops in Africa. In West Africa, it is also known as guineacorn, in the Sudan as durra, and in southern Africa as kafir. It was domesticated long ago in northeastern Africa.[11] It is an erect grass usually 1 to 4 meters tall. It grows well on heavy soils, but will also grow satisfactorily on lighter, sandier soils. It tolerates periods of drought and/or waterlogging better than maize, and similarly has more tolerance for soil acidity (being grown in soils of pH values ranging from 5.0 to 8.5) or salinity. Sorghum is often grown in mixtures with early or late millets, groundnuts, cowpeas, and cotton, and often in 2- to 5-crop mixtures. Red sorghum is usually grown near compounds, white sorghum in outfields.[12]

One noteworthy feature of sorghum varieties is that by means of photoperiodism the date of heading is closely related to the average date of the end of the rains so that seed set occurs as the weather becomes dry, avoiding problems of mould and insect attack. This feature, however, imposes a rather narrow north-south range of adaptability. Date of sowing has a marked effect on yields because grain is formed largely on moisture stored in the soil. Thus, if rains end early or the crop has been sown late, grain fails to fill completely.

Sorghum is consumed as a thin porridge or thick paste, or as beer. White or yellow corneous grains are preferred for flour, which is prepared by pounding in a mortar to produce a fine powder. Stalks are used as building material, fuel, and feed.

Millet. Millets (different varieties of which are pearl millet (also known as bulrush millet) and finger millet, and which are referred to in French generally as le petit mil) were also domesticated in Africa and today are almost as widespread as sorghum.

Millet varieties cover a wide range of growing seasons in the SAT zone, from 55 to 180 days or more. Some millet varieties are cultivated in West Africa as far north as the 200 millimeter isohyet, where neither maize nor sorghum can be grown as rainfed crops. Millets thrive on light soils, and are well adapted to high temperatures and high solar radiation, and can exhibit very high rates of growth and water use efficiency. Their rate of root development is very high. They cannot, however, tolerate waterlogged conditions. They are fairly tolerant of soil acidity. Millets are less susceptible than maize and sorghum to stem-borers and weeds such as striga, but they are vulnerable to bird depradations.

Millets form the basis of many cropping mixtures and are often grown in outfields, but can be an infield crop as in the Sérèr pombod (see below). They were the staple of the Azande at the time the farming of these people was studied by De Schlippe. They are consumed as a porridge (in West Africa: <u>to</u>) after grinding, and as a beer. Stalks are used as building material, fuel, and feed.

<u>Maize</u>. Maize is a fairly recent arrival in Africa. Growing season varies from 80 to 140 days according to variety. It prefers well-drained loam soils. Its pH tolerance ranges from 5.0 to 8.0. On soils of low fertility, however, maize will produce little or no yield, unlike sorghum and millet. High maize yields are not possible without adequate root nutrition. It has less tolerance to drought than millet and sorghum due to its more superficial root system. It may be grown as a sole crop as well as in mixtures, but maize intercropping rates are commonly as high as 84 percent or 75 percent of total area of the crop.[13] It is intercropped with a wide variety of crops, such as cereals (including upland rice), legumes, vegetables, root crops, bananas, and cotton. Sometimes in a complex crop mixture the maize looks like no more than a stalk here and there rather than what would be described as a maize field.

Maize is usually grown near compounds so as to benefit from the fertilizing effects of household wastes. It has the great advantage over sorghum and millet that its husk protects the ear from bird and insect depradations and from damage by rain during the ripening period. Consumption is in the form of a thick paste from ground flour in the SAT and green on the cob as a vegetable in the TRF.

<u>Rice</u>. Portères in his seminal article identified two early centers of diffusion of the African rice <u>Oryza glaberrima</u> in the inland delta of the Niger River and on the upper Casamance and Gambia rivers.[14] He dates rice cultivation in Africa to 1500 BC. The Asian rice <u>Oryza sativa</u> entered Africa much later, being introduced by Arab traders to the east coast between the 8th and 10th centuries, by the Portuguese to the west and east coasts between the 15th and 16th centuries, and to Madagascar by seafarers across the Indian Ocean in the 16th century. Asian rice spread rapidly and now is by far the more important.

Rice is one of the most adaptable crops known, so it is grown both in the TRF and the SAT under appropriate conditions. Sowing of rice takes place in a variety of ways: e.g. dibbling seed into a dryland seedbed, broadcasting pregerminated seed in mud, and sowing into a nursery bed with the highly controlled water availability that defines true irrigation. Rice is grown in mangrove

swamps on the West African coast, in floodplains and
inland valleys benefiting from impounded runoff and from
seepage, and along seasonally flooded river banks as far
north as the desert edge. Heavy alluvial soils are better
suited than lighter soils. Optimum pH appears to be about
7.0, but upland rice will grow well at 5.0 to 6.0, and on
black alkaline soils rice grows at 8.0 to 9.0. Although
there is some irrigated rice grown in Africa, the proportion is minuscule compared to the portion grown without
complete water control (table 1.1).

Fonio. This crop is considered to be the oldest
indigenous African cereal and is still an important food
staple in the Sahel of West Africa.

Root Crops. Yams have been grown in west-central
Africa for 5,000 years and formed the basis of what was
probably the continent's earliest agricultural system. As
Coursey states, "The domestication of the yam in Africa
may thus be viewed as an essentially indigenous process
based on wild African species."[15]

TABLE 1.1
Types and Area of Rice Cultivation, West Africa, 1976

Code	Type	Percentage of total area
	Total	100.0
1	Upland rice cultivation	65.0
11	Strictly upland cultivation	62.5
111	Hill rice	5.0
112	Flatland rice	57.5
12	Groundwater cultivation with rains	2.0
13	Groundwater cultivation without rains	0.5
2	Lowland rice cultivation	35.0
21	Mangrove rice cultivation	8.0
211	Without tidal control	2.0
212	With tidal control	6.0
22	Freshwater cultivation	27.0
221	Without water control	22.0
222	With partial water control	3.0
223	With complete water control	2.0

Source: WARDA, "Classification of Types of Rice
Cultivation in West Africa," mimeo, 1978, p. 20.

Yams are planted in hillocks or mounds which are between 0.6 and 1.3 meters high and 0.9 to 1.3 meters apart. They propagate by cut pieces of tubers known as setts. A frequent practice is for farmers to plant yams on newly cleared land because yams, unlike cassava, perform poorly on soils cropped in the previous season. They may be intercropped with millets, sorghum, maize, rice, cowpeas, groundnuts, sweet potatoes, and cassava.

Cassava arrived in Africa in the 16th century, originally in the TRF, later in the SAT. Growing period ranges from 9 to 24 months, with good conservation in the ground. This feature makes cassava adaptable to a wide range of cultivation frequencies, from long fallow to continuous cultivation. Cassava is grown either in sole stands or mixed with other crops. Crops grown together with cassava commonly include plantains and beans or sweet potato, and maize and groundnut or beans in East Africa; and yams or yams and maize, or more complex yam mixtures in West Africa.

Cassava is the principal staple of the Luba in the Congo Basin and other peoples. Among the Banda, for instance, cassava displaced sorghum as the principal crop.

Other important root crops in Africa are potatoes, sweet potatoes, and cocoyams.

Legumes and Oil-Bearing Crops. Crop legumes include groundnuts (peanuts), pigeonpea, common beans, and cowpeas, and constitute important sources of protein, having high complementarity with cereals, in African diets. They are important agronomically because of their soil nitrogen fixing properties which enhance soil fertility.

Cowpeas were probably domesticated in Ethiopia. They are rarely grown as a sole crop. Groundnuts are consumed as stews and roasted or boiled, and their oil is used in cooking. Crop residues are used as fuel and feed. Pigeonpea is intercropped with maize, sorghum, or sweet potato. In East Africa, dried pigeonpeas are mixed with cassava and boiled in a mash known as kimanja.

Bambara groundnuts (voandzou) and Kersting's groundnuts are pulses that have been found wild in West Africa, where they must have been domesticated. The former is now found spasmodically throughout tropical Africa and is of some importance in Zambia.

Oil palm is undoubtedly of West African origin, where it occurs wild in riverine forests and freshwater swamps. It cannot survive in dense or primeval forests. Shea butter is an important oil crop of the West African SAT.

Bananas and Plantains. These are tall, rhizomatous perennials. They require rainfall in excess of 1,250 millimeters per year and well-drained soils. They are tolerant of acid soils with pH of 3.4. They are often integrated into annual cultivation, where they can provide shade needed by certain crops such as cocoa.

Bananas and plantains were the staple food of several TRF zone societies early in the 20th century in the extensive literature surveyed by Miracle (Rega, Yombe, Lese, northern Mongo, Bali, Budu, etc.).

Crop Yields

African farmers generally obtain yields that are far lower than yields of similar crops in temperate climates. Their yields are low even in comparison with such crops grown in Africa under different conditions (table 1.2). Abalu defines potential yields as being "levels that have been achieved without much difficulty under controlled conditions in all countries considered." The question is: Why don't African farmers get similar yields, assuming they are rationally motivated? What is the explanation of this apparently large "yield gap"?

TABLE 1.2
Selected African Potential Crop Yields

Crop	Yield (tons/ha)	
	(1)	(2)
Sorghum	4	4 (Northern Cameroon)
Millet	3	3 (Senegal)
Maize	6	10 (Madagascar, Ivory Coast)
Rainfed paddy	4	7 (Western Cameroon)
Wheat	3	
Roots & tubers	30	

Sources: Column 1: G. O. I. Abalu, Solving Africa's Food Problem," Food Policy, Aug. 1982, table 2, p. 249; column 2: C. Pieri, "Fertilisation des Cultures Vivrières et Fertilité des Sols en Agriculture Paysanne Subsaharienne," L'Agronomie Tropicale, 41, 1 (Jan.-Mar. 1986), p. 2.

The point is, African farmers do not grow their crops under what plant scientists refer to as controlled conditions, meaning that measures are taken to prevent pests, diseases, and other natural hazards from interfering with the yield experiments. African farmers farm in difficult natural conditions. To say, as agronomists quite properly do, that farmers in western Nigeria obtain low yields of maize because over half the land surface has a gravel horizon or stone line at a shallow depth, that total porosity and water holding capacity of the soil are low, making for low root elongation rates and root volumes, thus rendering the plants susceptible to drought,[16] is to say nothing more than that African farmers face a difficult task to extract crops from their land. This is a statement economists should be comfortable with. Instead of seeking to change the environment (which is in any case impossible in most cases) or recommending new breeding programs for higher-yielding crop varieties, economists should think about ways in which productivity rates can be improved in the context of existing land and other resources.

For African low-resource farmers, traditional varieties with their relatively low yields seem to be preferred over high-yielding varieties known to perform well under controlled conditions, but more vulnerable to diseases and pests than the traditional varieties, less competitive with weeds, and (most importantly) requiring inputs unobtainable or management practices not easily duplicated by farmers. The principle of revealed preference is one familiar to economists. A rational explanation can usually be found for such preferences.

There is, moreover, a serious flaw in most published comparisons of African crop yields. African crops are usually grown in mixtures. What is often measured as "crop yield" in Africa is the output of a crop of interest per unit of cultivated land in isolation from the other crops which may be growing on the same unit of land. It is hardly surprising that yields of crops like maize and sorghum in Africa are low when compared with yields of these crops obtained in American or European agriculture, or under irrigated conditions in Africa, since a mixed crop is being compared with a sole crop. Yet even the low yields measured on this basis in Africa are not inferior to yields measured for the same crops grown under similar conditions in South Asia (table 1.3).

It does not follow automatically, then, that only substitution of "exotic" crop varieties having quite different physiological characteristics from traditional varieties, together with packages of inputs, or of an "exotic" tool or machine for traditional cultivating implements, will raise productivity in low-resource agriculture in Sub-Saharan Africa. At the very least, change of this type, to be applicable on a wide scale in low-resource agriculture in Sub-Saharan Africa, would have to meet the minimum requirement of compatibility with mixed cropping. This is a very large order, and suggests the thought that research on ways to improve other productivity-related aspects of African agriculture than yield may more readily produce results.[17]

TABLE 1.3
Comparative Crop Yields

Region/country	Yield (kg/ha)	
	Sorghum	Millet
Africa:		
Nigeria	633	620
Sudan	700	433
All	696	594[a]
South Asia:		
India	675	561
All	683	558

[a]West Africa only.
Source: David J. Andrews, "Current Results and Prospects in Sorghum and Pearl Millet Breeding," paper presented at the 6th World Bank Agriculture Sector Symposium, Development of Rainfed Agriculture under Arid and Semi-Arid Conditions, Washington, D.C., January 1986, tables 1 and 2.

TOOLS

African farmers' tools are extremely simple and are made of wood, iron, and other readily available materials. Simplicity of design and manufacture, however, does not preclude diversity. The array of tools used for preparing seedbed, planting, cultivating, and making earthworks like ridges and channels attests to a long history of experimentation, borrowing, and adaptation.[18] Adoption rates of animal traction in areas where it has not been traditional have generally been low, however.[19]

A close symbiosis exists between tools and crops. Some generalization is possible about tools and their uses, as about crops and theirs:

1. A single village usually contains a large tool inventory.
2. Tools vary by their degree of aptitude for soil conservancy tasks.
3. Tools vary by labor time required per unit of land.
4. Tools have specialized uses, as between men and women for instance, explicable in anthropological, historical, or other terms.[20]

Crops and tools constitute, between them, a large share of the working capital of low-resource agriculture.

NOTES

1. Paul W. Richards, "The Tropical Rain Forest," Scientific American, Vol. 229, No. 6 (December 1973), pp. 58-67.

2. J. M. Kowal and A. H. Kassam, Agricultural Ecology of Savanna: A Study of West Africa (Oxford: Oxford University Press, 1978).

3. Presentation by Claude Charreau at the 6th World Bank Agriculture Sector Symposium, Development of Rainfed Agriculture under Arid and Semi-Arid Conditions, Washington, D.C., January 1986.

4. The single most useful survey of food crops in lowland tropical Africa I have found is the volume of collected papers presented at seminars at the International

Institute of Tropical Agriculture sponsored jointly by the Ford Foundation and the Institut de Recherches Agronomiques Tropicales et des Cultures Vivrières edited by C. L. A. Leakey and J. B. Wills, Food Crops of the Lowland Tropics (Oxford: Oxford University Press, 1977). Also useful is M. J. T. Norman, C. J. Pearson, and P. G. E. Searle, The Ecology of Tropical Food Crops (Cambridge: Cambridge University Press, 1984). A. H. Kassam, Crops of the West African Semi-Arid Tropics (Hyderabad: ICRISAT, 1976) includes non-food crops.

5. Guillemin cites the example of 18 genotypes of sorghum grown mixed in the same fields by farmers on the Oubangui Plateau. (R. Guillemin, "Evolution de l'Agriculture Autochtone dans les Savannes de l'Oubangui," L'Agronomie Tropicale, XI, 2 (1956), p. 167.)

6. Peter R. Jennings, "The Amplification of Agricultural Production," Scientific American, September 1976, pp. 181-183.

7. Miracle documents the introduction of "exotic" crops in the Congo basin by Portuguese and Arab traders from the 15th century. (Marvin P. Miracle, Agriculture in the Congo Basin (Madison: University of Wisconsin Press, 1967), pp. 231-244.) Maize, cassava, sweet potatoes, potatoes, beans, tomatoes, and chillies are all "recent" introductions to Sub-Saharan Africa. Bananas, cocoa, tea, and rubber are also "exotic" to Sub-Saharan Africa.

The development and diffusion of packages of high-yielding cotton varieties in the francophone countries of West and Central Africa is the outstanding example today of successful technological innovation and adoption among African low-resource farmers. The following data covering 11 countries are cited by Charreau (loc. cit.):

	1973-74 average	1983-84 average
Rainfed cotton production (tons)	286,000	470,000
Average cottonseed yields (kg/ha)	547	1,061
Percent of total area fertilized	34	76
Average fertilizer dosage (kg/ha)	43	123
Average insecticide dosage (lt/ha)	4	9.2

It should be noted that cotton is monocropped in the SAT of Africa, and that the high-yielding varieties have been adopted by farmers with relatively little dislocation. The original introduction of cotton to African farmers occurred much earlier, in the colonial period, often by force, and caused considerable dislocation. (See, inter alia, Jean Cabot and Christian Bouquet, Le Tchad (Paris: Presses Universitaires de France, "Que Sais-Je?" No. 1531, 1973), pp. 81–86.)

8. Peter J. Matlon and Dunstan S. Spencer, "Increasing Food Production in Sub-Saharan Africa: Environmental Problems and Inadequate Technological Solutions," American Journal of Agricultural Economics, 66, 5 (December 1984), p. 674.

9. Dunstan S. C. Spencer, "A Research Strategy to Develop Appropriate Agricultural Technologies for Small Farm Development in Sub-Saharan Africa," chapter in Ohm and Nagy (eds.), Appropriate Technologies, p. 315.

10. Peter J. Matlon, "A Critical Review of Objectives, Methods, and Progress to Date in Sorghum and Millet Improvement: A Case Study of ICRISAT/Burkina Faso," chapter in Ohm and Nagy (eds.), Appropriate Technologies, p. 174. ICRISAT is the acronym of the International Crops Research Institute for the Semi-Arid Tropics, headquartered in Andhra Pradesh, India.

11. J. R. Harlan, "Agricultural Origins: Centers and Non-Centers," Science, 174 (1971), pp. 468–474.

12. Information about the scientific descriptors of African crops presented here is largely based on Kowal and Kassam, Agricultural Ecology of Savanna, and Norman, Pearson, and Searle, Ecology of Tropical Food Crops, and on Jack R. Harlan, Jan M. J. De Wet, and Ann B. L. Stemler (eds.), Origins of African Plant Domestication (The Hague: Mouton Publishers, 1976).

13. Norman, Pearson, and Searle, Ecology of Tropical Food Crops, p. 116.

14. Roland Portères, "Vieilles Agricultures de l'Afrique Intertropicale; Centres d'Origine et de Diversification Variétale Primaire et Berceaux d'Agriculture Antérieurs au XVIè Siècle," L'Agronomie Tropicale, Vol. V, Nos. 9–10 (1950), pp. 489–507.

15. D. G. Coursey, "The Origins and Domestication of Yams in Africa," chapter in Harlan, De Wet, and Stemler (eds.), Origins, p. 403.

16. Norman, Pearson, and Searle, Ecology of Tropical Food Crops, p. 111.

17. Compare the following thoughtful statement:

The obsession with yield which most agricultural specialists from the developed world or from Asia bring to Africa is as counter-productive in projects as it is in research.

(Hans Binswanger, "Evaluating Research System Performance and Targeting Research in Land-Abundant Areas of Sub-Saharan Africa," Discussion Paper No. 31 (Washington, D.C.: World Bank, January 1985, mimeo), p. 16.)

18. African farm tools have been catalogued in ORSTOM, Les Instruments Aratoires en Afrique Tropicale (Paris: 1984).

19. The spotty record of animal traction in Sub-Saharan Africa has been reviewed by Yves Bigot, "Quelques Aspects Historiques des Echecs et des Succès de l'Introduction et du Développement de la Traction Animale en Afrique Sub-Saharienne," Machinisme Agricole Tropical, No. 91 (1985), pp. 4-10.

20. The diffusion both of crops and tools across Africa was, of course, a far more complicated process than is suggested here. For an interesting attempt to relate ethnographic distribution of crops in one region, and to present a tentative periodization of the establishment of African crops, see Nicholas David, "History of Crops and Peoples in North Cameroon to A.D. 1900," chapter in Harlan, De Wet, and Stemler (eds.), Origins, pp. 223-267.

2

Production

Traditional agriculture in Africa suffers from the image of being backward. A principal feature associated with backwardness of this sort is its "subsistence" orientation. African agriculture, it is supposed, produces barely enough for the needs of producers. There is no surplus.

The evidence, however, both historical and present-day, is otherwise. Commodity trade has a long tradition throughout the continent, indicating the presence in historical times of some form of surplus. For present-day conditions, calculations made by Richards, on the basis of a variety of field studies from both the SAT and TRF zones, show to what extent African agricultural systems based on traditional crop varieties and hand hoes are capable of producing a surplus (table 2.1).

MANAGEMENT PRACTICES

The management of physical factors during the cropping season is generally more difficult in tropical agriculture than in the temperate zone. Routines derived from precise knowledge of the environment are needed for successful agriculture because alternatives as to use of factors of production are fewer than in environments where residual soil moisture is more easily controlled.

The necessary knowledge for such management practices may have been acquired by generations of farming experience, or by scientific observation and deduction, or a combination of these. There have been very few studies by agronomists, however, of the relationship between African farmers' management practices and the sustainability of agricultural production, perhaps because such studies require long periods of observation.

Crop yields in Africa depend, speaking strictly in agronomic terms, on the balance of soil moisture and the balance of soil nutrients. The meaning of this observation in terms of human behavior is that the farmer's skill in managing these two balances can directly affect yield.

In the SAT zone, potential evapotranspiration is higher than rainfall for most of the year. This means that the farmer has to exploit the additions to soil moisture during the brief rainy season in order to produce a crop.

TABLE 2.1
Energy-Based Input-Output Ratios for Some African Food Production Systems

Food production system	Ratio[a]	Observations
Upland rice, Sierra Leone	1:9	Includes intercrops
Swamp rice, Sierra Leone	1:5	
Sorghum, northern Nigeria	1:9	Range 1:5 to 1:16
Sorghum/millet, northern Nigeria	1:7	Range 1:6 to 1:9
Yam/cassava, eastern Nigeria[b]	1:8	Infields 1:11, outfields 1:6
Yam/cassava, eastern Nigeria[c]	1:5	Infields 1:7, outfields 1:2
Yam, central Nigeria	1:7	Average, valley and upland fields
Hunter/gatherer, Kalahari	1:8	
Pastoralists, Uganda	1:5	

[a]Defined as number of days' worth of food earned by one day's labor.
[b]Medium population density.
[c]High population density.
Source: Paul Richards, "Ecological Change and the Politics of African Land Use," The African Studies Review, Vol. 26, No. 2 (June 1983), table 3, p. 30, which cites original data sources. Reprinted by permission.

Crops vary greatly in their cultivation requirements. Some can be sown with a minimum of field preparation. Others need a preparatory hoeing to loosen the soil and uproot weeds (notably in the case of planting root crops like cassava). Therefore, we would expect that labor requirements vary by crop. Norman and co-workers have gathered data on labor requirements of different crops grown in the Hausa area of northern Nigeria. These data are summarized in table 2.2.

The time dimension is of paramount importance in determining both inputs and outputs in low-resource agriculture in Sub-Saharan Africa. Delayed planting often results in a reduction in the effective length of the growing season and lower yield. Delays in first weeding of growing crops lead to lower yields. Soil moisture balance depends also on the crops grown, since some are more water-demanding than others. As Kowal and Kassam have summed up the situation, "A precise time-tabling of farming operations imposed by the water regime is of paramount importance for the timely and successful preparation of land and for the establishment of crops, weeding, the efficient use of fertilizers and for productivity."[1]

What crops and varieties the farmer plants, and when and how, will be largely influenced by his or her expectations as to the weather. This makes it all the more astonishing that African farmers are able to "outperform" the weather. The reliability of the output of African agriculture, even where rainfall is extremely unreliable, has been well documented. Collinson provides data covering three consecutive years in three similar areas in Sukumaland (Tanzania) showing a level of staple grain supply much more stable than the amount of critical rainfall and more stable on farmers' fields than on a trial farm unit rated at a very high level of management (table 2.3). These findings substantiate the frequent observation that traditional African crop varieties are yield-reliable, but they also say much about farmers' skills.

Management of Soil Fertility

This means that the farmer has an interest in maximizing the number of variables in his or her production process--land type, crop variety, planting

TABLE 2.2
Labor Input Differentiation by Crops Grown, Sokoto,
Northern Nigeria

Crop mixture	Sample size (number of plots)	Average labor input[a] (man-hours/ acre)
Upland crops:		
Millet/cowpeas/red sorrel	6	76.37
Groundnuts	7	153.56
Millet/late millet/sorghum/cowpeas	6	153.86
Millet	6	163.94
Millet/sorghum/cowpeas/red sorrel	30	170.74
Millet/cowpeas	61	185.60
Millet/groundnuts/cowpeas	5	190.25
Millet/sorghum	7	204.45
Sorghum	7	210.21
Millet/sorghum/cowpeas	127	225.98
Millet/sorghum/groundnuts/cowpeas	3	231.76
Sorghum/deccan hemp	7	252.71
Cassava	29	353.42
Lowland crops:		
Rice, cassava/calabash (double cropped)	7	225.72
Rice	28	382.88
Rice, calabash (double cropped)	20	406.43
Rice, calabash/tobacco (double cropped)	7	614.69
Rice, cassava (double cropped)	7	790.32
Sugarcane	15	877.67

[a]All operations, including field preparation.
Source: D. W. Norman, J. C. Fine, A. D. Goddard, W. J. Kroeker, and D. H. Pryor, A Socio-Economic Survey of Three Villages in the Sokoto Close-Settled Zone, 3. Input-Output Study, Vol. 1, Text, Samaru Miscellaneous Paper 64, 1976, table 60, p. 91. Reprinted by permission.

timetable, harvesting timetable, and so forth. Nowhere is
this so important as with respect to soil fertility.

Throughout the production process, the farmer's skills
are oriented to conserving the fertility of the land base,
a need that is met, in the absence of available chemical
fertilizers on a large scale, by highly adaptive and
tested cultural practices. Poor soil fertility depresses
crop yields, other things remaining the same. But it also
exerts a variety of other more complex feedback effects,
such as restricting the choice among crops that can be
planted.[2]

Aside from depending on its moisture retention
capacity, the fertility level of a soil depends on its
content of nutrients available for plant growth. Loss of
such nutrients therefore occasions a lowering of the level
of fertility. Such loss can be caused by many factors,

TABLE 2.3
Stability of Production on Farms in Sukumaland

Item	Year		
	First	Second	Third
Data from farm surveys:			
Staple grain supply--			
Pounds per adult equivalent	392	368	324
Index (1962-64 = 100)	109	102	90
Rainfall--			
Amount (mm)	504	306	370
Index (1962-64 = 100)	130	78	94
Data from trial farm unit:			
Staple grain supply--			
Pounds per adult equivalent	385	215	750
Index (1963-65 = 100)	74	42	145
Rainfall--			
Amount (mm)	424	318	353
Index (1963-65 = 100)	116	87	96

Source: Michael Collinson, Farm Management in Peasant
Agriculture (Boulder, Colo.: Westview Press, 1983), table
3, p. 28.

and against these the farmer is in constant struggle in the effort to obtain a harvest.

The most obvious form of nutrient loss is that which occurs when vegetative matter is removed from the field and not returned. Mention has been made of the relatively small removal of soil nutrients by tree crops in the TRF zone by agricultural crops like coffee and cocoa. Other crops account for larger removals of soil nutrients (table 2.4), thereby challenging the farmer to find suitable ways to restore such nutrients for the succeeding crop or crops.

Other ways in which soil nutrients are lost to crops are through water and wind erosion, leaching, sun-baking and crusting, and by uptake by weeds.

Conservation of resources in African agriculture was originally achieved by use of the factor time through allowing land to lie fallow in order to restore soil fertility. Farmers also conserve resources by use of crop sequences or rotations, taking advantage of the beneficial effects some crops have on soil properties like mineral and organic matter content and physical structure in such a way as to counter the effects of other, more depletive crops. Finally, simultaneous mixed cropping is a way African farmers have long conserved resources.

TABLE 2.4
Nutrients Removed from Soil in Harvest
of Some African Crops

Crop	Yield (kg/ha)	Nutrients removed (kg/ha)				
		N	P	K	Ca	Mg
Maize (grain only)	1,100	17.1	3.0	3.0	0.2	0.2
Rice (paddy)	1,100	13.6	3.5	3.9	0.9	1.5
Groundnuts	770	30.7	2.6	5.3	1.0	
Cassava	11,000	25.0	3.3	66.0	5.9	
Yam	11,000	38.6	3.0	39.9	0.7	
Bananas	11,000	30.7	4.5	63.2	0.7	
Cocoa	11,000	25.0	4.4	36.4		

Source: FAO, Organic Recycling in Africa, Soils Bulletin No. 43 (Rome: 1980), p. 26, table 9.

FALLOWING

Crop rotation based on fallowing is a system in which successive crops on the same piece of land, rather than following one another on an annual basis, are interspersed with several years of fallow in which the land reverts temporarily to bush or forest. This reversion allows accumulation of vegetative matter, which restores the nutrients to the edaphic complex[3] through litter fall, rainwash, timber fall, root decomposition, and burning. In turn, this important process makes possible regeneration of the biomass, the total mass of living matter in the soil, both vegetable and animal, of which crops become a useful part.

A fallowing system is completely described by the length of its cycle and by the proportion of the cycle that is devoted to planting, tending, and harvesting crops. African low-resource agriculturalists are adept at managing these two dimensions of productivity in accordance with their environment.[4]

Farmers in the "yam belt" of central Nigeria illustrate the high degree of ability to manipulate cultivation practices to suit the environment. These farmers live in the southern SAT, in an area of hills and valleys with variable soil qualities. Yams are grown in various crop mixtures. Farmers in Tawari village generally follow a strict two-year crop rotation. The rotation is started with yam and relay intercropped with pearl millet and African yam bean. Very often farmers extend the plot for millet so that on some area around the yam, millet is grown as a sole crop. In the second year, farmers commonly plant cowpeas on half the plot and maize on the other half. Both crops are then relay intercropped with sorghum. Sometimes farmers grow cereals in sole stands. The breakdown of area planted in this village is 43 percent yam-based mixtures, 40 percent non-yam mixtures, and 17 percent sole cropping.

In Osara and Eganyi villages, crop rotations and combinations are not standardized, with 54 different combinations having been recorded in Osara not including minor vegetable crops. Yam intercropped with vegetables and early maize is usually followed by fallow or it is intercropped with cassava which is allowed to mature in the second year before the land is fallowed. Other combinations include melon/cowpea/maize/sorghum and melon/cowpea/cassava/sorghum.

Yams are known to make significant demands on soil fertility, and thus the latter factor becomes an important determinant of the dimensions of the fallowing system. At Tawari, on rather poor soils, farmers continuously and uniformly crop their land for about 15 years and then fallow it for 5-6 years. Farmers at Osara and Eganyi, on the other hand, use a short-cycle fallowing system on lowlands: crops are produced for 1-3 years and then fields are fallowed for about the same length of time. These farmers are thus able to maintain a fairly stable, high level equilibrium of soil fertility. On uplands in these same villages, however, crops are produced for 6-7 years running, followed by an equal time of fallow. Such relatively long cycles of cropping-fallowing are said to result in a low-level equilibrium with fluctuating soil fertility. The frequency of years with maximum fertility, that is years immediately after fallowing, is thus higher in the short lowland rotations than it is in the long upland rotations, meaning that yams can be grown more often in the short rotations of the lowlands.[5]

Cultivation Frequency

The duration of the cropping phase in a fallowing system as a percentage of the total cycle is called the cultivation frequency. This may be expressed mathematically, following Joosten:[6]

$$\text{Cultivation frequency} = \frac{\sum A_i C_i}{\sum A_i C_i + \sum A_i F_i}$$

where A_i is area of field i, C_i is the number of years of cultivation of field i since the last lengthy fallow, and F_i is the number of years of fallow since the beginning of the last lengthy fallow.

Figure 2.1 shows the fluctuations in residual edaphic fertility associated with a fallowing system with a cycle length of 15 years and a cultivation frequency of 20 percent, using diagrams which Guillemin, on the basis of his studies in Central Africa, called Van der Pool curves. The sharp drop in fertility due to cropping (shaded area) represents loss of soil nutrients in the offtaken crop.

The bottom part of figure 2.1 shows what happens to edaphic fertility if the cycle length is reduced (implying an increase in cultivation frequency) without any compensating action to add nutrients to the soil. Guillemin cites examples of this having happened on the Oubangui Plateau.

Generally, intervals between crop cultivation in a stable fallowing system vary according to the time

Figure 2.1 Fluctuations in Residual Edaphic Fertility under a Fallowing System

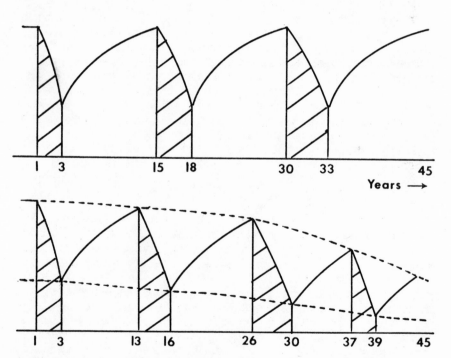

Vertical axis represents fertility of edaphic complex. Cross-hatching indicates cropping.

Source: Drawn on the basis of figure in R. Guillemin, "Evolution de l'Agriculture Autochtone dans les Savannes de l'Oubangui," L'Agronomie Tropicale, XI, 2 (1956), p. 158.

required for regeneration of the biomass. Note that there is not necessarily a unique relationship between cultivation frequency and residual soil fertility: the relatively infertile upland soils of Tawari have a cultivation frequency of one-half those of the relatively more fertile lowland soils of Osara and Eganyi.

At the end of the fallow period, the native or adventist species that have invaded the plot are burned off and crop planting again takes place among the charred stumps of trees. This burning process is an essential part of resource conservation, for it adds minerals to the soil at the same time as it clears the way for cultivation operations. Often there are successive burnings: the first a few weeks after trees have been felled by hand or killed by ringing, the second after an initial planting in order to reduce tree trunks to ashes for their mineral content, and a third more complete burning of piled-up debris to form the planting ground for crop varieties which do especially well on high mineral content. Without burning, soil mineralization would proceed too slowly to be of use for agriculture. Clearing land by burning is often called wasteful for its failure to regenerate organic matter in the soil, but this can be counted as one of the costs of this particular cropping system, much like gasoline as one of the costs of mechanized cultivation of wheat.

Mobility of Inputs

From the point of view of mobility of inputs, it is obvious that land use under this system is strictly governed by the amount of allowable fallow time. Burning requires relatively little labor use, and labor use is thus concentrated in planting and harvesting operations. Sowing and planting are usually done directly in the ashes without any other land preparation and by using a digging stick. For practitioners of fallowing, "no-till" farming is nothing new. If the allowable fallow time is reduced, because of the need to feed a larger population from the same land base (governed perhaps by walking distance from the compound), then means must be found to prevent the rapid loss of soil fertility caused by shortening the fallow period illustrated by the bottom half of figure 2.1.

Where the forest cover is thinned by burning and large trees no longer have time to take root and grow before the next burning, crop yields can only be preserved by adding minerals from an outside source, as is necessary in the SAT zone. Burned or unburned leaves or other vegetable materials or turf are sometimes brought to the cultivated plots and mixed with the topsoil by means of hoeing. Additional fertilization is provided by animal and human manures and by kitchen wastes. Under more intensive cultivation, several of these methods of fertilization are used simultaneously.

As the fallow period is shortened, more sunlight reaches the ground and grasses invade land that was formerly under forest. This evolution poses a different problem to cultivators: burning is no longer an efficient way of clearing plots for cultivation because grass roots are left intact by burning and grasses regenerate quickly, crowding out planted crops.[7] Also, there are infestations of weeds, which are totally lacking in the tropical rain forest. Therefore, some mechanical means of soil tillage is required: usually the hoe is the first solution to this problem, and may give way at a later stage to the plow. Use of the plow in fields recently reclaimed from forest requires destumping, the most labor-intensive of all operations connected with African agriculture.

MIXED CROPPING

Mixed cropping refers to any method by which more than one crop is grown on a given land area at any one time. Mixed cropping is the general rule in African low-resource agriculture. Numerous studies have shown that except for some small areas in East and Southern Africa, mixed cropping is vastly more important than sole cropping systems, occupying over 90 percent of cropped area in most countries.[8] The benefits, both direct and indirect, are numerous, and African low-resource agriculture encompasses a wide array of variations within the method.

Further categorizations of mixed cropping may be:

Intercropping--More than one crop on a given area at one time arranged in a geometric pattern.
Relay cropping--A form of intercropping where not all the crops are planted at the same time.
Sequential cropping--More than one crop (or inter-crop) on a given area in the same year, the second crop being planted after the first is harvested.

Because crop mixtures are so widespread among African societies with a corresponding variability in the rules that appear to govern the ways in which the crops are combined or alternated, the usual practice is to refer to crop associations, crop sequences, or simply crop mixtures as in this book, rather than to crop rotations. Certain characteristic crop mixtures can be recognized as recurring among certain people in specific environments, as is the case with crop rotations in Western agriculture, but the variability extends well beyond natural environments to encompass even such short-term influences as rainfall in the course of the growing season.

Mixed cropping may be described in terms of three dimensions: (1) crop composition; (2) spatial organization; and (3) timing.

There is good evidence that mixed cropping is a technically efficient way of using resources for production.[9] The combinations and permutations of crops in African mixed cropping systems, within the constraints imposed by the natural environment, are very numerous. The number of possible crop combinations is greatest in the TRF zone, where up to 60 crop species are sometimes counted on a single farm, although 20-30 is believed to be nearer the norm. The combinations adapted to forest fallow are particularly ingenious, with shade-tolerant species being planted on the margins of the plot, moisture-demanding species at the bottom of slopes, nutrient-demanding species on localized ash concentrations or on hoed-up mounds of topsoil, and climbing species against unfelled tree trunks or next to crops with rigid stalks like maize, and so on. In the SAT, 10 to 15 crop species are normally found on a single farm.[10]

Spatial organization of mixed cropping is an important way of obtaining technical efficiencies in using resources for production. Some crop combinations take advantage of the differential root densities and soil penetration of the crops in the mixture so as to exploit soil nutrients and moisture. Others benefit from shading. Often there exist synergetic relationships between the mixed crops. A frequent motive for mixed cropping in the SAT is that it helps to spread the risk of crop failure. Farmers explain that in case of drought they stand a chance of harvesting one or more drought-resistant crops from a field even if others in the same field fail.

Within the limits of the space of their own farms, African farmers seem to try to engage in agricultural production on as wide a set of soil and microclimatic conditions as possible. Richards describes farms in central Sierra Leone (TRF) that typically have a more extensively cultivated upland portion and a more intensively cultivated valley-bottom "wet land" portion to take advantage of micro-differences favorable to their crops which are well known to the farmers.[11] In the SAT, extensively-cultivated outfields and intensively-farmed infields commonly constitute many farms. Farmers in long-settled SAT areas deliberately fragment their cropping patterns in order to offset microclimatic variations and to localize pests and disease attacks.[12] Communal African systems of land tenure obviously facilitate such diversification of resource use.

Timing in mixed cropping systems may be thought of in terms of the crop-specific variables of planting date, maturity period, and harvest date. From the farmer's point of view, examples of desirable goals in selecting these variables may be differentiating nutrient demands on the soil of various crops in the mixture at various stages of growth, evening out labor demand for cropping operations like harvesting, reducing storage losses by keeping crops in the ground as long as possible, and assuring a supply of harvested food throughout the year.

From the resource conservation point of view, one of the major benefits of mixed cropping is that it keeps a vegetation cover on the soil. Clearing land for cropping greatly increases its susceptibility to soil erosion (table 2.5),[13] and also accelerates leaching of soil nutrients and exposes the soil to sun-baking. These considerations explain why African farmers try to keep a vegetation cover on their fields for as long a portion of the rainy season as possible, and also explain the frequent choice of tree crops like bananas or coffee in cropping patterns. Total biomass production with mixed cropping is higher than with sole cropping and more residual biomass is returned to the topsoil.

Looking at crop output, the yield, nutritional, and economic efficiencies of mixed cropping have been shown on investigation. Taking results achieved on the principal crops alone, maize yields measured in the Ijenda region are higher in several of the crop mixtures than in single-crop stands. Similar results have been measured in the SAT zone. Singh, for instance, found that farmers' yields of both millet and sorghum in three areas of Burkina Faso were higher in mixed cropping with cowpeas than in single-cropped fields, of which there were (understandably) few.[14]

Furthermore, the output achieved on additional secondary crops is often impressive. Johnny, on the basis of a sample of 31 upland farms in southern Sierra Leone having an average of 21 intercrop species per farm plot, calculates that secondary crops (rice being the principal crop) accounted for 53 percent of the market value and 73 percent of the food energy value of the total output of those farms.[15]

The calculations in table 2.6, on the basis of data derived from a farm survey in what is now Zaire, provide evidence that the method of mixed cropping is more efficient than that of sole cropping both in terms of protein production and in terms of economic return. One hectare of mixed cropping produces more protein than one hectare of any of the individual crops sole-cropped (col. 4), and larger economic return than any of the individual crops except cassava and sweet potato (col. 5).

The data in column 6 of table 2.6 show the land use efficiency of mixed cropping. Summing the land areas required in sole cropping (B) to produce the quantities

TABLE 2.5
Susceptibility of Soil to Erosion, Sefa, Senegal
(15-Year Averages)

| Land cover | Water from rainfall (percent) | | Erosion (tons/ha/year) |
	Runoff	Infiltration	
Forest	1.0	99.0	0.2
Fallow	16.1	83.4	4.9
Crops	21.1	78.8	7.3
None (bare soil)	39.5	60.5	21.3

Source: Claude Charreau, "The Technology: Is It Available? Where Is It Being Developed?," paper presented at the 6th World Bank Agriculture Sector Symposium, Development of Rainfed Agriculture under Arid and Semi-Arid Conditions, Washington, D.C., January 1986, p. 3.

(weight) grown on a single hectare of mixed cropping (A), we see that it would have required 55 percent more land in aggregate. Put another way, the land equivalent ratio in the case of this crop mix was 1.55.[16]

Because of the very heavy weeding-labor requirements of sole cropping, this would be a disadvantageous system compared with mixed cropping from the farmer's point of view, even if land were in abundant supply. In the words of some farmers in the TRF of Cameroon who tried it unsuccessfully, "Monocropping is too much work for too little output."[17]

Finally, it should be pointed out that the figures in col. 1 (B) show <u>estimated potential</u> yields of the same crops grown under similar conditions. But if farmers were to attempt to cultivate these crops in sole stands, they might expose the soil to water and wind erosion and, without additional fertilizer from outside the system, might provoke a permanent decline in soil fertility. With continuous cropping, even the application of inorganic fertilizer alone would in all likelihood not prevent a permanent decline in soil fertility such as could only be prevented by the application of manure. In other words, mixed cropping is a way to maintain relatively good crop yields without damaging the resource base. We may say, therefore, that the Van der Pool curves in figure 1 are in the long term a specific function of the cropping mix and associated cultivating practices.[18]

Abalu and D'Silva have proved the rationality of mixed cropping for achieving targets of minimizing risk and meeting household nutritional requirements with the use of a linear programming model using data for a SAT area in northern Nigeria.[19]

The economic efficiency of an agricultural system is measured by the degree to which the marginal value products of resources used in production approximate the cost of those resources. One of the well-known difficulties of investigations of low-resource or subsistence agricultural systems is the absence of markets for many resources such as land and labor.

Prudencio, in an intensive study of a mixed-farming village in the SAT zone of West Africa, devised a method of avoiding having to find market prices to value inputs land, hand-tool labor, animal-traction labor, organic fertilizer, and mineral fertilizer, which certainly have non-zero shadow prices, in circumstances where these markets are extremely thin or non-existent.

TABLE 2.6
Comparative Efficiencies of Mixed and Single Cropping

Cropping method and crop	Average yield (kg/ha)	Protein content (percent)	Unit value[a] (frs/kg)	Protein production (kg/ha)	Economic return[b] (frs/ha)	Land equivalent ratio[c]
	(1)	(2)	(3)	(4)	(5)	(6)
Mixed cropped (A):						
Maize[d]	86	10.0	3.50	8.60	301	
Sorghum[d]	565	10.4	8.00	58.76	4,520	
Cassava[e]	1,157	0.7	2.00	8.10	2,314	
Sweet potato[e]	2,428	1.5	3.00	36.42	7,284	
Beans[d]	377	24.0	7.00	90.48	2,639	
Total				202.36	17,058	1.00
Single cropped (B):						
Maize[d]	1,500	10.0	3.50	150.00	5,250	0.06
Sorghum[d]	1,200	10.4	8.00	124.80	9,600	0.47
Cassava[e]	12,000	0.7	2.00	84.00	24,000	0.10
Sweet potato[e]	8,500	1.5	3.00	127.50	25,500	0.29
Beans[d]	600	24.0	7.00	144.00	4,200	0.63
Total						1.55

[a]Market prices collected in 1956 and 1957 in current francs.
[b]Column 1 multiplied by column 3.
[c]Number of hectares required to produce quantities in column 1 (A) of crops, either in mixture or in single cropping.
[d]Dry grain.
[e]Fresh roots.
Source: Columns 1, 3, and 5: J. Hecq, "Le Système de Culture des Bashi (Kivu, Territoire de Kabare) et Ses Possibilités," Bulletin Agricole du Congo Belge, XLIX, 4 (August 1958), pp. 992–993; column 2: FAO, Human Nutrition in Tropical Africa (Rome: 1965), appendix table 3; columns 4 and 6: calculated by the author.

The method depends on calculating values for the shadow prices of these inputs by using as proxies the marginal value products resulting from fitting production functions to the entire set of input and output data collected. Prudencio's main objective was to reveal differences in input productivities among different cultivation practice- and crop-related subsamples. The subsamples for which Prudencio collected data and fitted production functions (of Cobb-Douglas and linear forms) were five different types of fields (identified by their location with respect to the compounds and the crops grown on them) and three specific crop mixes. Prudencio recognized explicitly that farmer management varied systematically in accordance with these field types and crop associations.

Accordingly, Prudencio computed values of a productivity ratio, r, based on the following formula:

$$r_j = \overline{Y}_j / \sum \overline{MVP}_i \overline{X}_{ij}$$

where $\overline{Y}_j$ represents the mean value of output in the jth subsample, $\overline{MVP}_i$ represents the marginal value product of input X_i at its mean level of use during the year of observation, and $\overline{X}_{ij}$ represents the mean amount of input X_i used in the jth subsample. The denominator of r_j stands for the mean opportunity cost of the major inputs used in the following production operations in the subsample: seedbed preparation, weeding, and late ridging.

In other words, the ratio r is an economic input-output ratio and is as close a measure of the economic efficiency of an African low-resource agricultural system as we are likely to find. The values of r, calculated using different functional forms, are given in table 2.7.

Prudencio gives the significance of his calculations in the following statement:

> Given the above definition of r, the difference between the numerator and the denominator of r may be viewed as an estimate of returns to management and to resources used during harvest and processing. Since r itself is a benefit-cost ratio, it may be viewed as a simple measure of efficiency within the restricted analytical space where only output prices and crop production possibilities matter.[20]

Flexibility of Mixed Cropping Systems

A basic difference between African low-resource agriculture and commercialized agriculture is the flexibility of the mixed cropping systems that characterize the former. As Steiner describes it:

> This means that the farmer does not know exactly what he will grow on his fields in the middle or at the end of the season. The final cropping pattern depends on the time between the onset of the rains and the latest possible planting dates for individual crops, on intervening drought periods, and on the availability of labor:

TABLE 2.7
Productivity Ratios in Nonghin Village (Burkina Faso), 1981

Subsample	r [a] (1)	r (2)	r (3)
Cropping rings:[b]			
Ring 1	0.39	1.77	4.29
Ring 2	0.98	1.13	1.94
Ring 3	1.07	1.27	2.60
Ring 4	0.76	0.73	1.64
Ring 5	0.80	0.76	1.24
Cropping mixture:			
Maize/sauce plants	0.30	1.59	3.50
Red sorghum/cowpeas	1.08	1.18	1.98
Millet/white sorghum/cowpea	0.64	0.62	1.16

[a]Prudencio states that excessively high marginal value products for fertilizer in the Cobb-Douglas production functions he estimated resulted in probably unrealistically low values for r in this series.
[b]Cropping rings are at greater distances from compound according to increasing numerical order.
Source: Yves Coffi Prudencio, "A Village Study of Soil Fertility Management and Food Crop Production in Upper Volta - Technical and Economic Analysis," unpublished Ph.D. dissertation, University of Arizona, 1983, table 39, p. 251. Reprinted by permission.

--If the rains are late, there is often not enough time to plant all crops as it will be too late for some of them, e.g. photoperiod-sensitive varieties (such as many cowpea varieties).

--Crops that have failed because of drought periods have to be replaced by others.

--The available labor depends on several factors, such as the number of active household members present during the planting period, the health of the farmer and his family, and the money available for hired labor.[21]

This situation is at its most extreme in the SAT zone, where the shortness of the growing season and the variability of rainfall impose a severe burden on managerial skills. A relative abundance of labor, for example, may allow farmers to make an early start on weeding their red sorghum crop, enabling them to raise yields and reduce total cultivated area. Farmers who fall behind in weeding (early rains may induce heavy weed growth, which is in any case rapid during June and July), on the other hand, face an uphill battle to control weeds throughout the remainder of the growing season as labor requirements for weeding per hectare increase exponentially because of the greater physical effort of removing large weeds. Consequently, they may prefer to save on labor by sacrificing yields and extending their cropped area.

Farmers who intercrop cowpeas with red sorghum may need to allocate labor for protecting their later-maturing cowpeas from the depredations of roaming livestock, while farmers who intercrop cowpeas with white sorghum or millet, which are later maturing than red sorghum, will find their cowpeas less vulnerable to grazing animals. In all these cases, farmers are substituting inputs in production within the production period, as the shadow prices of inputs like land and labor fluctuate due to weather, bottlenecks, and other causes.

Fresco observed the case of a widowed farmer in a dry area of the Kwango-Kwilu who planted mainly cassava, intercropped with patches of squash, on her slope field. When the cassava produced a poor harvest of roots and leaves, she decided to intercrop it with a few sweet potato plants. A second case involved a farmer who established two fields, one on the slope and another on the top of the plateau, as well as a small homestead garden. After four months it appeared that the cassava on the plateau field was heavily infested with bacterial blight and cassava mealybug. As she expected almost no return from that field, she then decided not to do any weeding and limited her work to the harvesting of <u>voandzou</u> and a few minor crops. Instead, she concentrated her work on the slope field, where she repaired the ridges that had been damaged by a rain storm.[22]

This sort of input and output substitution during the course of the production period would be difficult to imagine in commercialized agriculture, where production is programmed in advance in such a way that the farmer's defenses against adverse weather and disease do not include altering the crops grown. While African households have preferences among foods, these do not seem to impose rigid adherence to a particular cropping pattern.[23]

In African farming communities, the association among land, climate, crops, and tools on the one hand and management of resources on the other is such an intimate one that these communities have all developed characteristic nomenclatures of field types. The Congo Sundi, for example, speak in terms of a six-tiered system of fields in which the <u>musitu</u> (forest field on a slope where men and women crop cassava, vegetables, maize, bananas, and other crops) is readily distinguished from the <u>nseké</u> (savanna field where women grow groundnuts). Practices of fallowing, burning, cultivating, and harvesting are similarly specific to field type.[24]

Under these field-specific management systems fall the techniques that African low-resource farmers use to preserve their resource base while carrying on production: systematic use of fallow, adjusting cropping sequences, engaging in mixed cropping, retaining flexibility of decision-making with respect to crops grown, and investing labor in erosion control works and addition to the soil of organic manure and other materials. These may be termed management practices or applications. The methods themselves have been evaluated as being efficient in terms of comparative volume of output and nutritional adequacy of output in relation to other methods, and in the use of economically costly resources.

NOTES

1. Kowal and Kassam, Agricultural Ecology of Savanna, p. 94.

2. It is interesting that low soil fertility is cited in most discussions of African agriculture as a reason for low crop yields, but almost never as a reason for low mobility of factors among alternative crop outputs, which in the long run may have as much to do with limiting aggregate crop production. See, for example, Christopher L. Delgado and John W. Mellor, "A Structural View of Policy Issues in African Agricultural Development," American Journal of Agricultural Economics, 66, 5 (December 1984), pp. 665-670.

3. The soil with its physical and chemical properties together with its content of organic matter and moisture. The term edaphic community is used in this sense by John Phillips, Agriculture and Ecology in Africa (New York: Praeger, 1960), passim.

4. The scientific term "rotational bush fallowing" applied to this system was apparently first popularized by O. T. Faulkner, an agriculturalist with Indian experience, who was appointed director of the Nigerian Department of Agriculture in 1920. Faulkner was suspicious of the penchant of the colonial authorities of his day for offering African farmers untried innovations and considered their fallowing system to be rational, efficient, and stable. (See Paul Richards, Indigenous Agricultural Revolution; Ecology and Food Production in West Africa (London: Hutchinson, 1985), p. 20.)

Accounts of fallowing systems in Africa are given by William Allan (The African Husbandman (Edinburgh: Oliver and Boyd, 1965)), who calls fallow the chief protective device against erosion in African land-use systems (p. 386); by H. Vine ("Developments in the Study of Shifting Agriculture in Tropical Africa," in R. P. Moss (ed.), The Soil Resources of Tropical Africa (Cambridge: Cambridge University Press, 1968), pp. 89-119); by W. B. Morgan ("Peasant Agriculture in Tropical Africa," in M. F. Thomas and G. W. Whittington (eds.), Environment and Land Use in Africa (London: Methuen, 1969), pp. 241-272); by P. H. Nye and D. J. Greenland in "The Soil Under Shifting Cultivation," Commonwealth Bureau of Soils Technical Communication, No. 51 (Farnham Royal: Commonwealth Agricultural Bureaux, 1961); and by the Food and Agriculture Organization of the United Nations (FAO) (Shifting

46

Cultivation and Soil Conservation in Africa, Soils
Bulletin 24 (Rome: 1974)).

5. L. Diehl and F. E. Winch, "Yam Based Farming
Systems in the Southern Guinea Savannah of Nigeria,
Discussion paper No. 1/79, International Institute of
Tropical Agriculture (Ibadan, Nigeria), 1979, pp. 22-25.

6. J. H. L. Joosten, "Wirtschaftliche und Agrar-
politische Aspekte Tropischer Landbausysteme" (Goettingen:
Institut fuer Landwirtschaftliche Betriebslehre, mimeo,
1962).

7. See photo number 32 in Hans Ruthenberg, Farming
Systems in the Tropics (Oxford: Oxford University Press,
3rd ed., 1980), following p. 202.

8. Spencer, "A Research Strategy," p. 314.

9. Good discussions of the efficiencies of mixed
cropping in the African context are to be found in Kurt G.
Steiner, Intercropping in Tropical Smallholder Agriculture
with Special Reference to West Africa (Eschborn: Deutsche
Gesellschaft fuer Technische Zusammenarbeit (GTZ), 2nd
ed., 1984), pp. 63-82; and Hugues Dupriez, Paysans
d'Afrique Noire (Paris: L'Harmattan, 2nd ed., 1982), pp.
141-146. Buntjer has described how Hausa farmers shift
crops and ridges in their fields from one year to the next
so as to obtain maximum fertility effects. (B.J. Buntjer,
"Aspects of the Hausa System of Cultivation Around Zaria,"
Samaru Agricultural Newsletter, Vol. 13, No. 2 (April
1971), pp. 18-20.)

10. Norman found as many as 147 distinct mixed
cropping combinations in three villages in the Zaria
region of northern Nigeria. (David W. Norman, "Crop
Mixtures under Indigenous Conditions in the Northern Part
of Nigeria," Samaru Research Bulletin, 205 (1974).)

11. Richards, Indigenous Agricultural Revolution, pp.
86-106.

12. H. A. Luning, "Patterns of Choice Behaviour on
Peasant Farms in Northern Nigeria," Netherlands Journal of
Agricultural Science, XV (1967).

13. See the literature cited in R. Lal, "The Soil and
Water Conservation Problem in Africa: Ecological
Differences and Management Problems," in D. J. Greenland
and R. Lal (eds.), Soil Conservation and Management in the
Humid Tropics (New York: John Wiley and Sons, 1977), pp.
143-149.

14. Ram D. Singh, "Small Farm Production Systems and
Their Relevance to Research and Development: Lessons from
Upper Volta, W. Africa," mimeo, 1981, p. 63.

15. Michael Johnny, "Traditional Farmers' Perceptions of Farming and Farming Problems in the Moyamba Area," unpublished MA thesis, University of Sierra Leone, 1979.

16. In Mead and Willey's formulation, the result in column 6 of table 2.6 would have been arrived at by summing the ratios of yields in column 1:

$$\text{LER} = \frac{\text{Maize (A)}}{\text{Maize (B)}} + \frac{\text{Sorghum (A)}}{\text{Sorghum (B)}} + \frac{\text{Cassava (A)}}{\text{Cassava (B)}} +$$

$$\frac{\text{Sweet potato (A)}}{\text{Sweet potato (B)}} + \frac{\text{Beans (A)}}{\text{Beans (B)}}$$

(R. Mead and R. W. Willey, "The Concept of a 'Land Equivalent Ratio' and Advantages in Yields from Intercropping," Experimental Agriculture (Cambridge, U.K.), Vol. 16 (1980), pp. 217–228. For a further methodological discussion, see Willey, "Evaluation and Presentation of Intercropping Advantages," in the same journal, Vol. 21, No. 2 (1985), pp. 119–133.)

17. Alain Leplaideur, Les Systèmes Agricoles en Zone Forestière: Les Paysans du Centre et du Sud Cameroun (Montpellier: IRAT-CIRAD, 1985), p. 136.

Compare Steiner's statement based on interviews:

Given the constraints imposed by the limited supply of family or hired labor to overcome labor bottlenecks, farmers consider intercropping a practice enabling them to obtain higher returns both per unit of land and per unit of labor.

(Op. cit., p. 188.) Steiner maintains his observations apply to both the SAT and the TRF zones, and for varying densities of population.

For identical reasons, farmers in pre-World War II southern Nigeria rejected efforts to have them replace their indigenous seed cotton variety, which they intercropped, with American cotton, grown in monocropping, despite the more attractive price of the latter. Later, however, when they were offered the opportunity to plant an improved indigenous cotton variety, which they could plant with their other crops in the old way, so that it entailed no extra expenditure of labor, they adopted it with remarkable celerity. (O. T. Faulkner and J. R. Mackie, West African Agriculture (Cambridge: Cambridge University Press, 1933), pp. 10–11.)

Theoretical aspects of the labor efficiency of mixed cropping will be discussed below, pp. 69 et seq.

18. As De Schlippe observed:

> The shifting cultivator does not judge the value of his fallows by the duration of rest they have undergone, as does the European farmer. He has not kept a record of years. He knows the fertility of a piece of land, and its usefulness for one or another of his crops, from the type of vegetation which covers it.

(De Schlippe, Shifting Cultivation in Africa, p. 37.)

19. G. O. I. Abalu and Brian D'Silva, "Socioeconomic Aspects of Existing Farming Systems and Practices in Northern Nigeria," in ICRISAT, Socioeconomic Constraints to Development of Semi-Arid Tropical Agriculture (Hyderabad, 1979), pp. 3-10. See also the pertinent remarks made by Hans Ruthenberg in the same volume, pp. 37-39.

20. Yves Coffi Prudencio, "A Village Study of Soil Fertility Management and Food Crop Production in Upper Volta - Technical and Economic Analysis," unpublished Ph.D. dissertation, University of Arizona, 1983, p. 249. Harvesting and processing were relatively minor operations in Prudencio's study, meaning that the values of r reflect largely returns to management.

21. Steiner, op. cit., p. 209.

22. Louise O. Fresco, Cassava in Shifting Cultivation; A Systems Approach to Agricultural Technology Development in Africa (Amsterdam: Royal Tropical Institute, 1986), pp. 127-128.

23. Subject to other considerations, of course, like home consumption needs. Surprisingly, however, farmers' consumption preferences seem not to represent a great obstacle to such switches. See Michael Collinson, "Eastern and Southern Africa," chapter in Mellor, Delgado, and Blackie (eds.), Accelerating Food Production, pp. 86-87.

24. Dominique Desjeux, Stratégies Paysannes en Afrique Noire (Paris: L'Harmattan, 1987), pp. 158-161.

Catalogues of field types can be found in most serious studies of African farming communities in many different disciplines. Since in African low-resource agriculture management is inseparable from field type and crop association, it will not be treated as a separate input of production in the model presented below.

PART TWO

Dynamics and Change

3

Intensification of Production

From the comparative statics of farmers' management practices, I turn now to the dynamics of change in low-resource agricultural systems. Historically, as populations have grown and agricultural producers have devised new ways of using resources to provide food and income, relationships between inputs and outputs have changed. This process goes under the general term intensification. This chapter explores how this process works in practice.

"NEW HUSBANDRY" AS A CASE OF INTENSIFICATION

The English agricultural revolution of the 18th century is a case in point. Livestock occupied a central place in the English farming system, as is the case with African low-resource agriculture today. The English agricultural revolution occurred in the face of growth rates of population which alarmed some observers, and involved the replacement of the open field system with its one-third of arable land left fallow by an intensive rotation of arable land between food grains and feed crops. The new rotation supported larger livestock numbers, which in turn provided manure for maintaining soil fertility. Land productivity increased dramatically as a result, labor productivity less so because of the heavy labor requirements of turnip cultivation. The new system provided food for the rapidly rising population from which the labor force for the industrial revolution was recruited. The English agricultural revolution, in other words, did not involve new technology, only "new husbandry."[1]

Intensification will be defined more rigorously in terms of economic theory below. Generally speaking, however, African agricultural production under growing population pressure may be viewed as a constant struggle to prevent a fall in both soil fertility and crop yields as fallow periods are shortened and resources are used more intensively.[2] The search for new ways of combining crops and animals is an integral part of this process. Since there is a serious danger of permanent damage to soil fertility and structure, the maintenance of consumable output per hectare, not to mention consumable output per man-hour, as intensification occurs must be considered a not insignificant achievement on the part of African farmers.[3]

Although evidence from cross-sectional samples cannot be expected to substitute for historical time series of data, a sample of 340 farming households in 9 villages in northern Nigeria appears to show a more intensive resort to application of organic manure in an effort to prevent soil fertility from falling when fallow periods have been reduced due to population pressure (table 3.1).[4]

TABLE 3.1
Correlation between Fallow Period and Use of Manure

Item	Study area		
	Bauchi	Zaria	Sokoto
Population density (pers/km^2)[a]	24	31	49
Land fallowed (percent)	31.3	16.8	3.3
Cattle ownership (percent)[b]	12.7	15.8	38.2
Organic manure applied (tonnes/ha)	0.53	2.71	3.71

[a]Average for the province.
[b]Percent of families owning cattle.
Source: Compiled from data in David W. Norman, David H. Pryor, and Christopher J. N. Gibbs, "Technical Change and the Small Farmer in Hausaland, Northern Nigeria," African Rural Economy Paper No. 21, Michigan State University, 1979.

Crop yields may be preserved as a result of new fertilization practices, or may even increase somewhat. But the additional labor required increases at a much faster rate. Tasks such as artificial soil fertilization and tillage, weeding operations, and especially the production of fodder crops for draft animals in regions characterized by long dry seasons are very labor-consuming. As a result, labor productivity drops unless more advanced tools are brought to bear, and the introduction of these often coincides with the abandonment of the fallow period altogether and the adoption of continuous cultivation.

A high cultivation frequency is a hallmark of agricultural intensification. An East African society that had attained a cultivation frequency of 100 percent was described in some detail as early as the 1930s.[5] In point of fact, successful intensification of African agriculture can result in impressive input-output ratios without significant degradation of the resource base, as may be seen from the following examples.

AFRICAN CASES

The first system illustrated is the mixed farming system of the Sérèr people of the southern part of the so-called "groundnut basin" of Senegal, which has an average population density on the order of 75 persons per square kilometer. This is in the SAT zone. The Sérèr crop rotation pattern uses space (generally flat) to great advantage to maximize land and labor use during the relatively brief single rainy season when all crop cultivation must be performed. The system is diagrammed in figure 3.1, which shows one year's operations of crop rotation and livestock management.

The area around the Sérèr compound is known as the pombod. Because of fertilization by small animal manure and kitchen wastes, the pombod can be cropped continuously year after year. Here are grown a millet known as "souna" for its highly appreciated dietary quality, intercropped with cowpeas (which, being a legume, constitutes another soil fertility retention input). Beyond the pombod are the large fields devoted to sectors growing food grains and groundnuts. The basic food grain is another millet, "match" (sanio). Sorghum is a a complementary food grain whose four or five varieties are known collectively as "bassi" and are carefully chosen for soil suitability. The area sown to these food grains depends on the outlook for the harvest of souna, which is sown first in the year.

Crop rotation in the large fields follows a clockwise pattern by sector around the compound. Match and bassi are sown on land newly reclaimed from fallow. The relative shortness of the rainy season forces farmers to clear land before the advent of the rains so as not to lose precious time in the growing season; this practice, unfortunately, leaves the soil vulnerable to wind erosion. Groundnuts are sown on land previously in food grains.

Livestock raising in the Sérèr system is an integral necessity, as Pélissier remarks. At the beginning of the rainy (cropping) season, the larger livestock are grazed by day on fallow land (the fallow land of two adjoining communities being sometimes contiguous so as to minimize fencing). During the night livestock are tethered in the food grain sector on the plot to be sown, and tethered livestock can be rented out to neighbors who wish to fertilize their plots. As the large fields are harvested, however, livestock are put to graze in these fields during the day. Later, after the cowpea harvest, livestock are tethered in the pombod itself during the night all the length of the dry season.

An important role in this system is played by the acacia albida tree, of which there are many in the region. This tree has an inverted vegetative cycle: it is in leaf during the long dry season and loses its leaves at the beginning of the rainy season. It thus shades the soil during the heat of the dry season, preventing undue rise of soil temperature and reducing evapotranspiration. The acacia albida has a long taproot[6] which reaches moisture deep underground and brings to the surface nutrients valuable for other plant growth at the beginning of the rainy season. As a result, one can trace a fertility gradient from the base of the tree. Furthermore, its leaves and seeds furnish valuable grazing material for livestock.

The second illustration is the mixed farming system in the Ijenda region of Burundi, which has reached such an intensive stage due to very high population density (200 persons per square kilometer) that only 8 percent of the cropped land is in fallow at any one time. This system depends for fertility maintenance on animal manure, the production of which is reported to be the main reason for

Figure 3.1 Spatial (Horizontal) Organization in the Sérèr
Agricultural System

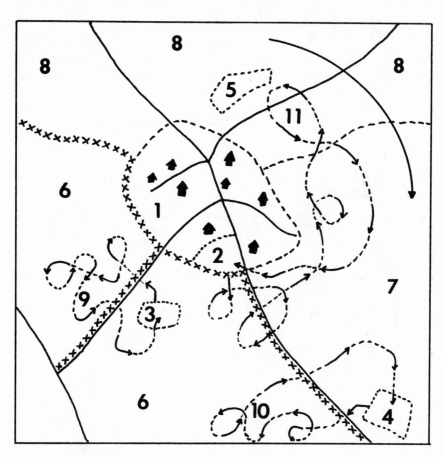

Legend: (Solid lines represent tracks. Crosses
represent thorn fences.) 1 Pombod. 2 Souna and cowpea
patch. 3 Future sorghum patch. 4 Present sorghum patch.
5 Present groundnut patch. 6 Fallow sector. 7 Match-
bassi sector. 8 Groundnut sector. 9-11 Livestock grazing
by day and tethering at night: 9 During April-September.
10 During September-December. 11 During December-April.

Source: Redrawn from Hugues Dupriez, Paysans d'Afrique
Noire (Paris: L'Harmattan, 2nd ed., 1982), p. 120, based
on information in chapter 5 of Paul Pélissier, Les Paysans
du Sénégal (Saint-Yrieix: Imprimerie Fabrègue, 1966).

cattle raising.[7] For protection of hillsides against erosion the system depends on either tree groves (the natural forest cover having disappeared long ago), crop cover, or ditching and terracing works.

This region, like most of Burundi, is extremely hilly. There are two rainy seasons, with the average annual rainfall of 1,550 millimeters being relatively well distributed throughout the year. The number of different crops cultivated is consequently very large, and their location within the farming unit depends to a great extent on altitude, slope, and microclimate (figure 3.2).

Figure 3.2 Spatial (Vertical) Organization in the Ijenda Region (Burundi)

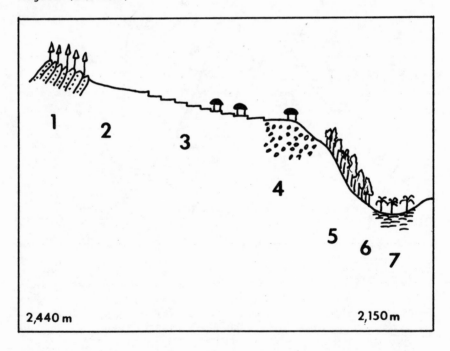

Legend: See opposite page.

Source: L. D'Haese and P. F. Ndimira, "Etude Multidisciplinaire des Systèmes d'Exploitation Agricole dans la Région d'Ijenda," Vol. 1, Bujumbura, 1985, p. 41. Reprinted by permission.

Agricultural operations are arranged so as to spread the use of land and labor throughout the year (figure 3.3). Actual measured crop yields obtained from this system are fairly high. The numbers in table 3.2, although they reflect different cultivating practices, especially organic manure applications, give some idea of yields obtained.

The crops grown in the Ijenda region are not limited to those in table 3.2. In Burundi agriculture, mixtures of sorghum, beans, and maize are common. In some cases, annual crops like beans are found on the same plots with crops with longer growing seasons like cassava and banana trees, or even with permanent tree crops like coffee or oil palm.

TABLE 3.2
Maize Yields, Single—Cropped and Intercropped,
Ijenda Region

Cropping pattern	Yield (kg/ha)		
	Sample area 1	Sample area 2	Sample area 3
Single cropped	1,400	2,500	1,925
Beans intercropped	1,187	2,200	2,780
Peas intercropped	1,228	1,980	2,000
Potatoes intercropped	1,912	2,800	3,180
Beans and potatoes intercr.	2,650	3,100	3,475
Beans and peas intercropped	1,138	2,280	2,422

Source: Compiled from D'Haese and Ndimira, "Etude Multidisciplinaire," pp. 161 and 165.

Legend to figure 3.2: 1 Groves of 3—year—old pines on loudetia simplex pasture. 2 Pasture of loudetia simplex on shallow soil with anti—erosion ditches. 3 Terraced fields of deep red soils for staple crops. 4 Granular lateritic soil planted to staple crops. 5 Eucalyptus grove. 6 Pasture with digitaria. 7 Swampland sometimes drained for maize and beans. Altitudes in meters a.s.l.

58

Figure 3.3 Cropping Pattern in the Ijenda Region

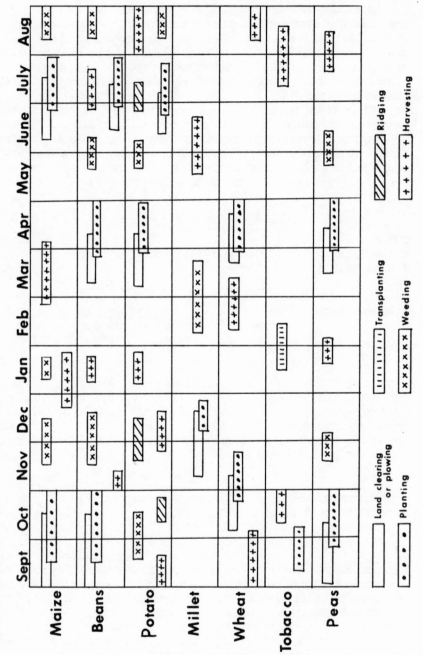

The detailed study of three villages in eastern Nigeria by Lagemann provides a third illustration of a traditional African society that has developed a highly intensified system of agricultural production using its own internal resources. Lagemann's study is particularly interesting from two points of view. In terms of human environment, the three villages illustrate population densities far in excess of even the Burundi case. And from the methodological point of view, Lagemann investigated correlation of other socioeconomic and agronomic variables (including soil fertility) with the differing population densities and farming practices in the three villages.

The three villages had population densities respectively of 250 persons per square kilometer, 500 persons per square kilometer, and 1,200 persons per square kilometer.[8] Practices such as manuring, mulching, spreading ashes, composting, and shifting latrines to make use of night soil were routinely followed in the high-population-density village, but much less so in the low-population-density village.[9]

Lagemann, from his field-by-field observations of soil fertility indicators, concluded as follows:

Two trends are apparent from the data on soil fertility:
(1) Within a village compound plots are more fertile than near fields, and near fields tend to be more fertile than distant fields. Soil fertility indicators improve with increasing land use intensity. This may be explained by fertilizing practices and by the effect of tree crops, where the root systems act as nutrient pumps.
(2) Between the villages there are pronounced differences in the fertility status of the soils. The higher population density and subsequently less fallowing result in lower levels of organic matter, nitrogen and phosphorus in the soils of the medium and high density villages. . . If soil fertility is to be maintained without fallowing, then counteracting efforts by the farmers have to be very intensive, comparable to those practiced in the compounds.[10]

INTENSIFICATION THROUGH APPLICATION OF
"MODERN" TECHNOLOGIES

Would intensification be possible through "modern"
technologies dependent on mineral fertilizers or mechanized sole cropping? The available evidence points to
the likelihood that mineral fertilizers or mechanization
would not furnish sustainable means for intensifying
low-resource agriculture in either the SAT or the TRF.
Results of a long-term IRAT experiment in which the
same plots of land in a SAT area of Burkina Faso have been
cultivated year after year since 1960 using the same
management practices lead to the conclusion that mineral
fertilizers are an inadequate, and possibly even detrimen-
tal, substitute for traditional methods of fertilization.
In this experiment, crop yields fell after the first few
years (figure 3.4).
Lest the assumption be made that modern varieties of
sorghum and legumes more responsive to fertilizer applications would prove the sustainability of the system where
traditional varieties do not, it is important to look at
what happened to the soil as well as to crop yields.
Trends in soil fertility indicators are documented in
table 3.3. Scientists reported a distinct depletion of
soil nutrients and an acidification harmful for crops.
Organic carbon in the soil depends heavily on the
respective rates of withdrawal and addition of organic
matter with fallowing or cropping on the soil. In the
long-term experiments at Saria, inorganic fertilizer made
little or no difference to the level of soil organic
carbon, whereas addition of manure had a marked impact on
the level of this nutrient. This may be seen in table 3.3
by comparing the time trend of the organic carbon level
resulting from annual applications of light mineral
fertilizer alone with the results of similar applications
of light mineral fertilizer plus manure. The time trend
results for heavy manure applications are even more
dramatic. Note that these experiments were designed to
measure only the chemical properties of the soil, and did
not measure the beneficial effects of additions of manure
on soil physical properties, such as porosity.
Nitrogen content of the soil is closely associated
with the soil organic matter, since it is almost all bound
in organic materials having been fixed once by the assimi-
lation of free nitrogen from the air by soil organisms.
Soils of the SAT are generally lower in nitrogen content

Figure 3.4 Yield Results of the Long-Term (1960-78) Fertility Maintenance Experiment with a Sorghum-Legume Rotation at Saria, Burkina Faso

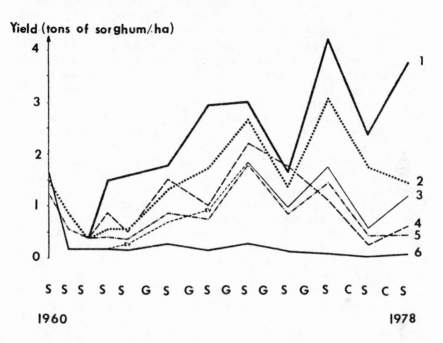

Legend: S Sorghum. G Groundnuts. C Cowpeas.
1 Heavy mineral (N-P-K) fertilizer applications and heavy manure applications.
2 Light mineral fertilizer applications and light manure applications.
3 Light mineral fertilizer applications plus plowing under crop residues and green manuring.
4 Heavy mineral fertilizer applications.
5 Light mineral fertilizer applications.
6 Control plot, no mineral fertilizers or manure.

Source: J. Pichot, M.P. Sedogo, J.F. Poulain, and J. Arrivets, "Evolution de la Fertilité d'un Sol Ferrugineux Tropical sous l'Influence de Fumures Minérales et Organiques," L'Agronomie Tropicale, 36, 2 (April-June 1981), chart 2, p. 125. Reprinted by permission.

than soils of the TRF. The experimental results for nitrogen in table 3.3 may have less direct bearing on soil fertility than those for organic carbon, since to be useful for crops soil nitrogen must first be converted to ammonium and nitrate forms. Nitrogen is lost to crops by burning, by leaching, and through competition from weeds. Nitrogen requirements vary greatly by crop and by stage of development. Kowal and Kassam cite experimental data showing that rice is about twice as efficient as millet and sorghum in terms of quantity of grain produced per unit quantity of nitrogen. Excessive nitrogen in cereals leads to lodging. In cotton, nitrogen application can increase pest infestation. These are problems pertinent to mixed cropping.

Phosphorus is next only to nitrogen in importance to crops in Africa. In the SAT, the generally small quantities of phosphorus in the soil and its tendency to react with soil components to form relatively insoluble compounds unavailable to crops results in widespread phosphate deficiency. Kowal and Kassam give an average phosphorus content of SAT soils ranging from 80 to 150 parts per million (substantially lower than for temperate regions). Continuous crop cultivation tends to deplete the soil of phosphorus, as may be seen from the control plot results in table 3.3. Even green manuring of crop residues does not prevent a lowering of the level of soil phosphorus because most of the plant's uptake of phosphorus goes to the grain. To avoid phosphorus from being a limiting factor in cropping, sizeable additions of phosphorus to the soil must be made. Even with heavy mineral fertilizer applications, however, manuring still makes a positive difference, as seen in table 3.3.

Kowal and Kassam give a usual range of 0.3 to 3.0 me of exchangeable calcium per 100 grams in SAT soils. They state that the calcium content of such soils tends to decrease significantly after mineral fertilizer additions due to leaching losses, the acidifying effect of the fertilizer, and removal through cropping. As a consequence, soil pH falls rapidly. In the experimental results in table 3.3 the application of heavy doses of mineral fertilizer has a worse effect on soil calcium availability than continuous cropping on the control plot. Addition of manure, however, counteracts this, leaving the soil exchangeable calcium level near the upper bound of the range given by Kowal and Kassam. This is reportedly due more to an indirect effect on the exchange

TABLE 3.3
Soil Fertility Indicators, Long-Term Experiment, Saria, Burkina Faso

Indicators and soil treatments	1969	1976	1978
Organic carbon (parts per 100):			
Control plot--			
Continuous sorghum	0.29	0.29	0.25
Sorghum-legume rotation		0.27	0.25
Light mineral fertilizer--[a]			
Continuous sorghum	0.29	0.30	0.24
Sorghum-legume rotation		0.31	0.26
Light mineral fertilizer plus green manuring--			
Continuous sorghum	0.31	0.33	0.25
Sorghum-legume rotation		0.35	0.28
Light mineral fertilizer plus light manure--			
Continuous sorghum	0.31	0.38	0.35
Sorghum-legume rotation		0.42	0.35
Heavy mineral fertilizer--[b]			
Continuous sorghum	0.31	0.36	0.24
Sorghum-legume rotation		0.35	0.28
Heavy mineral fertilizer plus heavy manure--			
Continuous sorghum	0.53	0.76	0.66
Sorghum-legume rotation		0.81	0.62
Nitrogen (parts per 1,000):			
Control plot--			
Continuous sorghum	0.23	0.30	0.18
Sorghum-legume rotation		0.24	0.24
Light mineral fertilizer--[a]			
Continuous sorghum	0.20	0.30	0.18
Sorghum-legume rotation		0.32	0.26
Light mineral fertilizer plus green manuring--			
Continuous sorghum	0.28	0.35	0.18
Sorghum-legume rotation		0.35	0.26
Light mineral fertilizer plus light manure--			
Continuous sorghum	0.29	0.40	0.44
Sorghum-legume rotation		0.43	0.32
Heavy mineral fertilizer--[b]			
Continuous sorghum	0.27	0.38	0.28
Sorghum-legume rotation		0.32	0.26
Heavy mineral fertilizer plus heavy manure--			
Continuous sorghum	0.47	0.77	0.54
Sorghum-legume rotation		0.79	0.63
Phosphorus (parts per 1,000,000):[c]			
Control plot	27.4	12.0/16.0	12.0
Light mineral fertilizer[a]	38.0	34.0/30.0	30.0
Light mineral fertilizer plus green manuring	31.7	28.0/28.0	24.0
Light mineral fertilizer plus light manure	34.8	36.0/41.0	33.0
Heavy mineral fertilizer[b]	42.4	43.0/53.0	38.0
Heavy mineral fertilizer plus heavy manure	70.7	59.0/87.0	59.0

See footnotes at end of table   --Continued

64

TABLE 3.3
Soil Fertility Indicators, Long-Term Experiment, Saria, Burkina Faso
(Continued)

Indicators and soil treatments	1969	1976	1978
Exchangeable calcium (me per 100 grams):			
Control plot--			
Continuous sorghum	0.82	0.74	1.15
Sorghum-legume rotation		1.05	0.99
Light mineral fertilizer--[a]			
Continuous sorghum	0.92	0.58	0.66
Sorghum-legume rotation		0.83	0.77
Light mineral fertilizer plus green manuring--			
Continuous sorghum	0.81	0.67	0.90
Sorghum-legume rotation		0.87	0.90
Light mineral fertilizer plus light manure--			
Continuous sorghum	0.98	0.82	1.14
Sorghum-legume rotation		0.87	0.90
Heavy mineral fertilizer--[b]			
Continuous sorghum	0.59	0.33	0.60
Sorghum-legume rotation		0.61	0.68
Heavy mineral fertilizer plus heavy manure--			
Continuous sorghum	1.75	2.25	2.37
Sorghum-legume rotation		2.43	2.62
Exchangeable potassium (me per 100 grams):			
Control plot--			
Continuous sorghum	0.20	0.17	0.16
Sorghum-legume rotation		0.17	0.12
Light mineral fertilizer--[a]			
Continuous sorghum	0.08	0.14	0.09
Sorghum-legume rotation		0.17	0.12
Light mineral fertilizer plus green manuring--			
Continuous sorghum	0.35	0.16	0.12
Sorghum-legume rotation		0.15	0.12
Light mineral fertilizer plus light manure--			
Continuous sorghum	0.21	0.21	0.22
Sorghum-legume rotation		0.26	0.22
Heavy mineral fertilizer--[b]			
Continuous sorghum	0.05	0.18	0.15
Sorghum-legume rotation		0.18	0.14
Heavy mineral fertilizer plus heavy manure--			
Continuous sorghum	0.60	0.61	0.50
Sorghum-legume rotation		0.62	0.65

See footnotes at end of table --Continued

TABLE 3.3
Soil Fertility Indicators, Long-Term Experiment, Saria, Burkina Faso
(Continued)

Indicators and soil treatments	1969	1976	1978
Soil pH:			
Control plot--			
Continuous sorghum	5.3	5.3	5.2
Sorghum-legume rotation		5.3	5.2
Light mineral fertilizer--[a]			
Continuous sorghum	5.1	4.9	4.6
Sorghum-legume rotation		5.0	4.7
Light mineral fertilizer plus green manuring--			
Continuous sorghum	5.1	4.8	4.7
Sorghum-legume rotation		4.9	4.8
Light mineral fertilizer plus light manure--			
Continuous sorghum	5.5	5.2	5.2
Sorghum-legume rotation		5.6	5.2
Heavy mineral fertilizer--[b]			
Continuous sorghum	4.7	4.5	4.4
Sorghum-legume rotation		4.7	4.5
Heavy mineral fertilizer plus heavy manure--			
Continuous sorghum	6.2	6.6	5.9
Sorghum-legume rotation		6.8	6.1

Note: Blank spaces indicate "Not measured."
[a]Mainly nitrogen and phosphorus.
[b]Nitrogen, phosphorus, and potassium.
[c]Phosphorus content of soil measured by Saunders method in 1969 and 1976, by Olsen-Dabin method in 1976 and 1978. All measurements were on continuous sorghum. No measurements were taken on sorghum-legume rotations.
Source: J. Pichot, M. P. Sedogo, J. F. Poulain, and J. Arrivets, "Evolution de la Fertilité d'un Sol Ferrugineux Tropical sous l'Influence de Fumures Minérales et Organiques," L'Agronomie Tropicale, 36, 2 (April-June 1981), tables 4 and 5, pp. 132-133. Annual doses of fertilizer nutrients and other detailed data on the experiments are provided in the original. Reprinted by permission.

capacity of colloids of the manure than to the calcium content of the manure.

Kowal and Kassam give a usual range of 0.18 to 0.25 me of exchangeable potassium per 100 grams in SAT soils. African crops are greedy for soil potassium, thereby severely limiting the amount of crop matter that can be taken off SAT soils without providing artificial fertilizer. In the experimental results in table 3.3, the beneficial effect of heavy mineral fertilizer plus manure is visible clearly.

The significance of soil pH to cropping is that it affects the availability of soil nutrients, it may inhibit desirable micro-organisms, and it may exert a toxic effect on crop root systems. Continuous cropping of sorghum or a sorghum-legume rotation in the absence of other farm management practices is enough to maintain a fairly stable value of soil pH, as seen from the control plot results in table 3.3. An increase in the amount of crop matter taken off the soil, however, results almost immediately in a fall in pH. The application of large amounts of fertilizer, particularly ammonium sulphate and urea, merely accelerates this process. In these conditions, uptake of phosphates is inhibited. The Saria experiments found a rise of aluminum toxicity with such treatment. The acidifying effect of heavy mineral fertilizer doses is evident from table 3.3.

If mineral fertilizers do not provide a feasible substitute for traditional fertilizing practices in Africa, do modern mechanical implements and power sources provide an alternative to the hand hoe?

Productivity gains by means of introduced technological change have been attempted in the SAT of Sub-Saharan Africa in the form of animal traction, the first step in mechanization of labor operations.[11] Theoretically, in the absence of constraints, replacing the hand hoe with the animal-drawn plow increases the marginal physical product of labor. Animal traction permits an expansion of area cultivated per person and, in conditions where land is in relatively plentiful supply, would appear to be an answer to breaking the labor bottleneck. But studies show that instead of breaking this bottleneck, it shifts it from some operations to others. Thus, animal traction in the SAT zone cuts seedbed preparation time, but increases weeding and harvesting labor requirements compared with use of hoes. Animal traction has been shown to require 10-14 additional man-days per hectare in groundnut or maize cultivation, 18-24 additional man-days in millet cultivation, and 24-32 additional man-days in cotton growing.[12]

In the mechanized Niger Agricultural Project started in northern Nigeria after World War II, it was originally expected that mechanical land preparation would make it possible for each farmer to crop 24 acres. This proved to be wholly impracticable, however, because without the mechanization of weeding, farmers simply could not cope with the task of weeding by hand, which required no less than 408 man-days of work over the six-week period.[13]

Soil erosion also poses obstacles to the introduction of mechanized sole cropping in Sub-Saharan Africa. The effort by Sodecoton in the SAT of Cameroon to transfer tractorized techniques of upland rice monoculture, for example, in the short space of seven years risks causing erosion damage in spite of the efforts of an army of extension workers. Warnings have already been given in this instance.[14]

There are thus serious obstacles to the introduction of mechanical technologies in Africa. Where such technologies are already "acclimatized," as in the case of animal traction in the Ethiopian highlands, it would seem that farmers were successful in preserving crop yields by adopting farming techniques compatible with the ecology of the region. The severity of Ethiopia's present erosion problem (due in part to cutting of trees for firewood) means that these same farming techniques have developed fallibilities. An interesting question is: Is reversion to a more "primitive" technology the only possible response to failures of this sort?

NOTES

1. See C. Peter Timmer, "The Turnip, the New Husbandry, and the English Agricultural Revolution," Quarterly Journal of Economics 83 (August 1969), pp. 375-395. I am not saying there exists an "evolutionary determinism" connected with this process. See the cautionary note in Richards, Indigenous Agricultural Revolution, pp. 139-140. Richards' entire book is an eloquent treatise on the innovative capacity of African farmers resulting from the self-adjusting character of low-resource agriculture rather than an argument for "evolutionary determinism." The two are quite distinct.

2. The relationship of population density to the intensity of agricultural production was examined in detail by Ester Boserup, whose book, The Conditions of Agricultural Growth The Economics of Agrarian Change under Population Pressure (Chicago: Aldine Publishing Co., 1965), was a landmark in the study of tropical agriculture.

3. See, for example, the calculations of factor productivities in table 5.2, below.

4. Compare the conclusion of Pingali, Bigot, and Binswanger: "The use of organic fertilizer is positively associated with the scarcity of land." (Prabhu Pingali, Yves Bigot, and Hans P. Binswanger, Agricultural Mechanization and the Evolution of Farming Systems in Sub-Saharan Africa (Baltimore: Johns Hopkins University Press, 1987), p. 5.

5. D. Thornton and N. V. Rounce, "Ukara Island and the Agricultural Practices of the Wakara," Monthly Letter, Department of Agriculture, Tanganyika, October 1933, revised edition, Nairobi (Kenya Pamphlet), 1945.

6. See the photo by Wickens of the unearthed root system of an accacia albida in E. A. Bell, "New Commercial Crops for Arid Areas," Journal of the Royal Society of Arts (London), July 1985, p. 547.

7. L. D'Haese and P. F. Ndimira, "Etude Multidisciplinaire des Systèmes d'Exploitation Agricole dans la Région d'Ijenda," Vol. 1, Bujumbura, 1985.

8. Johannes Lagemann, Traditional African Farming Systems in Eastern Nigeria; An Analysis of Reaction to Increasing Population Pressure (Munich: Weltforum Verlag, 1977), p. 178.

9. Ibid., pp. 38-39.

10. Ibid., pp. 52-54.

11. Animal traction in the TRF zone is limited by the tse-tse problem.

12. M. Le Moigne, "Animal-Draft Cultivation in Francophone Africa," Proceedings of the International Workshop on Socioeconomic Constraints to Development of Semi-Arid Tropical Agriculture (Hyderabad: ICRISAT, 1979), table 9, p. 218. See also the labor utilization graph reproduced in Steiner, Intercropping, p. 196.

13. K. D. S. Baldwin, The Niger Agricultural Project (Cambridge: Harvard University Press, 1957), p. 134.

14. P. Vernier, "Le Développement de la Riziculture sur Nappe Phréatique au Nord-Cameroun: Un Exemple de Transfert de Technologie," L'Agronomie Tropicale, 40, 4 (Oct.-Dec. 1985), pp. 323-336.

4

A Production Function of Low-Resource Agriculture

What follows is a model of African low-resource agriculture based on conservation of resources. The purpose of the model is to allow economic analysis of relations between inputs and outputs from the bottom up, and of factor productivities. I should warn the reader that, although the model contains some innovations in the way variables are defined, the analysis offers no radical departure from classical production theory.

Heady has placed farm conservation models in the more general setting of microeconomic theory:

> If we now turn to conservation from the standpoint of the individual farm, we find that the greatest number of farm practices and resource combinations which, in every day terminology, are called conservation investments are simply part of the general farm management problem and may or may not have important implications in use of soil resources over long periods of time. Rotations, liming and phosphate, or potash fertilizers fall in this category, as may contours or terraces employed for the purpose of capturing rainfall in moisture-deficit areas. The relationships and principles discussed previously with respect to crop combinations, livestock practices, and similar resource and product problems then serve as sufficient principles in outlining efficiency within this framework.[1]

It is suggested here that the production function of African low-resource agriculture is composed of two products and several inputs, one of which (labor) is allocable between the products annual output of crops and conservation of equilibrium biomass (CEB). The dependence of annual output of crops on CEB, and the reverse influence of annual crop output on CEB, assure jointness in the production function.

OUTPUTS

The product annual output of crops poses no particular problem other than measurement. Conceptually, it is the same as the classical production of agriculture the world over, in other words, the output of annual crops produced by combining land, labor, and capital. In low-resource agriculture, there is a serious problem of measurement, since many different crops or crop varieties are produced in mixture, so it becomes necessary to measure output by some index of weight or bulk, calories, protein, or cash value (market-determined or imputed).[2]

Conservation of equilibrium biomass as an output conveys the idea of the processes of crop and animal production in the SAT and TRF zones of Sub-Saharan Africa on the extremely variable (in both space and time dimensions) base of the edaphic complex. The main functions of these processes are restricting erosion, maintaining soil organic matter, maintaining mineral nutrient levels, avoiding the buildup of salts and acidity, and maintaining the availability and quality of soil moisture for plant growth.

The ways in which African low-resource farmers accomplish these conservation objectives are numerous and diverse, as we have seen. For model-building purposes, some of these are unmeasurable, like tethering cattle at night on plots intended to be cultivated, while others are measurable, such as hours of labor input into construction of bunds and field channels. But to be effective in producing annual crop output, the measures taken by farmers, be they measurable or unmeasurable, must be based on management skills. Having the right soils, crops, or tools is not enough: these must be combined so as to yield CEB, and thus sustain annual output. The production function portrays efficient production, and thus assumes a high degree of management skills on the part of farmers.

In the time dimension, CEB may be represented as a
line tangent to the Van der Pool curves in the top half of
figure 2.1, which as we have seen are a specific function
of the cropping mix and of the bundle of informed
cultivating and conservation practices applied to the
resource base. To provide a basis for sustaining annual
production, this line must be at least horizontal, or
rising, at a given value of the cycle length.

INPUTS

Labor is a variable input in African low-resource
agriculture, as it is in agriculture elsewhere. Rural
households, which supply the great preponderance of labor
used in agricultural production, contain a labor reserve
of family members that can be mobilized in case conditions
appear to the farmer highly favorable for crop and
livestock production. This, however, does not prevent
labor from being scarce at times of heavy labor demand, as
for weeding and harvesting. Labor can be diminished by
out-migration from the village to the city, or even
abroad, where it may be better paid.

Some labor in African agriculture is directed at
producing this year's crop. Such operations as sowing,
weeding, and harvesting are forms this type of labor input
takes. If one examines closely African farming systems,
however, another type of labor input appears. This type
of labor is associated with maintaining the operation of
the system through time, that is, with conserving equilibrium biomass. Operations such as clearing forest or bush
after fallowing, ridging, mounding, terracing, and digging
water channels are forms this type of labor takes.

The two types of labor are obviously complementary.
They are distinguished from one another primarily by the
way they bear on production. Labor in cropping has a
direct bearing on this year's output, since crop yields
are determined by such things as the amount of competition
from weeds for soil moisture and soil nutrients, and the
care with which the crop is harvested. Labor in land
preparation, on the other hand, has only an indirect
bearing on this year's output. The investment of labor in
field preparation maintains equilibrium biomass of the
soil for years to come. The absence of such labor
investment, implying as it does crop production of a
resource-exploitative nature, results in reduction of crop

yields starting this year or soon thereafter and, if the labor reduction is maintained, continuing into future years.

The cultivating practices and associated knowledge required to maintain equilibrium biomass at its necessary threshold cannot be changed in the short run (one cropping season or one fallowing cycle), and can only be changed in the long run. It takes years to restore equilibrium biomass by fallowing land, or by constructing anti-erosion terraces, for instance. Moreover, these practices must be learned. In the savanna areas of the Kwango-Kwilu, these techniques are used only by farmers with a long-standing tradition of cultivation, like the Bunda in southern Idiofa and northern Gungu. Farmers who have recently started cultivation in the savanna, on the other hand, hardly make use of any labor-intensive techniques such as mulching, composting, or ridging.[3]

In many African farming societies, the heavy sporadic labor involved in clearing "new" land for cultivation and building conservation works is associated with men, while the more continuous but equally heavy labor associated with hoeing, fertilizing, and harvesting crops is associated with women. The different purposes of these types of labor explains the fact that one often observes, in times of drought, men leaving the village to seek paying work elsewhere and the women who remain in the village laboring all the harder to help their crops survive the drought. There is no contradiction here. The labor of men and women in agricultural production is subject to a degree, and in some African societies a high degree, of specialization.

In many descriptions of African agriculture, the labor investment in conservation works and practices is completely ignored. Yet the explicit treatment as a distinct and fixed factor of production of the labor embodied in resource conservation avoids the erroneous argument that "natural fertility is doing most of the work that has to be done by human beings."[4] While it is true that crops grown in African fallowing systems benefit from the fertility stored up during the fallow years, the clearing of land anew is back-breaking work. Without it, the land would be useless for planting and harvesting. Such labor corresponds to the labor of spreading fertilizer in continuous cultivation systems.

Land area is also variable, as low-resource farmers increase or decrease the land area sown to crops as they

judge prospects to be more or less favorable or as their
needs vary for household consumption, community
obligations, and for replenishment of stocks.[5]

Variable capital consists of tools, seed, fertilizing
materials of all sorts, and crop varieties. All these are
available locally. Crop varieties are considered a
variable input because farmers commonly change their
intended cropping plans in accordance with short-term
forecasts of farming conditions: in the SAT zone if the
rains are late in coming, farmers sow millet in place of
sorghum, and so forth.

The notion of "intrinsic fertility," which is often
expressed in the literature,[6] is extremely important in
African low-resource agriculture. Attributes of intrinsic
fertility include the soil's organic carbon, nitrogen, and
mineral content, its physical structure and humus content,
moisture retention capacity and its variability, and other
natural features that make the soil apt for plant growth,
as well as land slope, microclimate, and other influences
that make the land vulnerable to erosion, leaching,
laterization, and other damaging processes. These are
long-term variables. Intrinsic soil fertility is known to
vary greatly within short distances, and of course can be
increased or decreased over time by management practices.
The concept of "intrinsic fertility" used here is thus not
one of "primordial" or "original" soil fertility, but one
of variable soil fertility.

Renewal of the biomass provides the nutrients for
plant growth, replacing those lost through water and wind
erosion and leaching, and the soil structure to support
plants, and that balances water exchanges with the atmo-
sphere, and transforms solar energy into plant material.
Practices facilitating renewal of the biomass include
measures to protect against diseases and pests and to
exploit synergetic biological relationships between different crops and between crops and livestock. Different
soils respond in varying degrees to such measures. Annual
rates of production of biomass vary among climatic zones
and by type of vegetation. Intrinsic fertility is
therefore an important, if long-term, factor determining
the conservation of equilibrium biomasss. Here, all
attributes of land other than surface area are encompassed
by the term intrinsic fertility and are considered to be
fixed for the duration of the production period.

Finally, cultivation frequency (previously defined in
the form of a percentage index), although fixed in the

short run, exerts a long-term stabilizing or destabilizing effect on equilibrium biomass, as we have seen.

The production function may be represented mathematically as follows:

$$y_1 = f(x_{11}, x_2, x_3; y_2) \tag{1}$$

$$y_2 = f(x_{12}, y_1; x_4, x_5) \tag{2}$$

where:

y_1 is an annual index of aggregate output of a specific crop mix,

y_2 is an annual index of CEB measured over several cropping seasons,

x_1 is labor input,

x_{11} is labor input into annual crop output,

x_{12} is labor input into CEB,

x_2 is land area input,

x_3 is variable capital input,

x_4 is a measure of "intrinsic fertility" of land, and

x_5 is the cultivation frequency.

Labor input is allocable between production of y_1 and y_2, subject to:

$$x_1 = x_{11} + x_{12} \tag{3}$$

There are a number of noteworthy features of the production function hypothesized in equations (1) and (2). The output y_2 is not merely a postponed output of y_1, as could be treated straightforwardly in a discounting model by determining farmers' time preference between rewards of production now and rewards postponed to later. In this hypothesis, the output y_2 depends on physical factors, notably the initial level of intrinsic fertility of land and the labor spent in deliberate efforts on the part of farmers to prevent degradation of the resource base. In other words, the farmer exercises some direct control over the time tradeoff: he or she is able to put off indefinitely the mortgaging of soil fertility by furnishing the necessary amount of labor in producing CEB.

What exactly is the relationship, then, between y_1 and y_2? The two outputs in equations (1) and (2) are organically linked. They don't just make use of some of the same inputs. In the terms defined in Carlson's 1930's dissertation, they are joint products.[7] This fact makes it impossible to ascertain even the technical relationships of inputs like labor and capital to one of these outputs without taking their relationship to the other into consideration simultaneously.

It follows from the form of the production function in equations (1) and (2) that the partial derivative of one of the outputs, say y_1, with respect to the other, y_2, expresses the technical substitution relation between the two outputs at a given level of variable inputs. In our case, the value of this derivative is the inverse of the partial derivative of y_2 with respect to y_1. That is, an increased quantity of one output may be obtained from a given quantity of variable inputs only if the quantity of the other output decreases.

Such a relationship, which holds in the short term, can be graphed by the product transformation curve in figure 4.1.

Figure 4.1 Product Transformation Curve

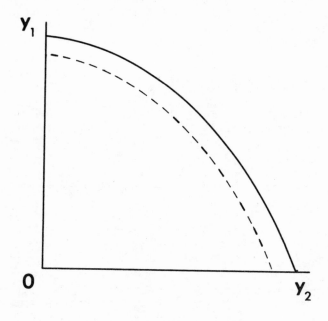

y_1 and y_2 do not figure as inputs in equations
(1) and (2) in the short term. Note that the relationships among variable inputs are likely to be positive in
these equations. In the long term, however, y_2 output
becomes an input into y_1 output, and vice versa. The
graphical treatment of outputs thus becomes more
complicated. Both y_1 and y_2 outputs can expand in
output space as all other inputs are increased, suggesting
a parallel outwards movement of the curve represented in
figure 4.1. But if y_1 expands to the point where CEB is
driven downwards, then the product transformation curve
might twist instead of moving in a parallel manner (more
y_1 and less y_2) in spite of a continued increase in
other inputs. The relationship between x_5 and y_2 in
equation (2) is a case of a negative interaction at all
possible levels on the basis of the previously discussed
relationship between cultivation frequency and soil
fertility.

Lastly, the function contains no explicit term for
technology. This is a deliberate omission, as will be
seen from the discussion which follows. Nevertheless, a
multiplicative technology parameter may be inserted, in
the usual way, as a term in equation (1) without loss of
generality if it is felt that such a parameter will
explain some significant share of the residual variability
not explained by the other terms in the equation.

INPUT SUBSTITUTION IN PRODUCTION

Before examining more closely productivities of
factors in low-resource African agriculture, we need to
extend our notion of technical change to accommodate
African realities. Technical change is defined as
substitution among inputs, implying in theoretical terms
movement along the production function constrained by
fixed inputs, y_2 in equation (1) in particular. By
definition, this production function is the technically
efficient resultant of inputs applied and outputs achieved
within one production period, that is to say one year.
Strictly speaking, the output y_1 represents a specific
crop mix.

Reducing the terms of equation (1) for illustrative
purposes to two inputs, labor and land area, technical
change may be represented in variable input space as
movement along an isoquant, or line of equal output, for

this particular bundle of output. In figure 4.2, for instance, land may be substituted for labor within the production period by moving along the isoquant (which is graphed only for the region where marginal products of both variable inputs are positive, and production therefore is rational) from point A to point B. This form of input substitution is a common tendency in Africa's land-extensive agricultural systems. A and B both show efficient use of resources in the production of y_1: they both represent the minimum quantities of those two inputs in different combinations that are necessary to produce a given quantity of y_1.

We are used to identifying y_1 with the duration of the production period, which we have defined in equation (1) as one year. Substitution of inputs, therefore, poses no great conceptual problem, since the isoquant remains unique for a well-defined production period. In African low-resource agriculture, however, farmers are faced with the need not only to adjust their cultivating practices, including input use, constantly within the production period, but also their crops in accordance with changing weather and resource availabilities, as we have seen.

Figure 4.2 Technical Change (Input Substitution, Production Function Fixed)

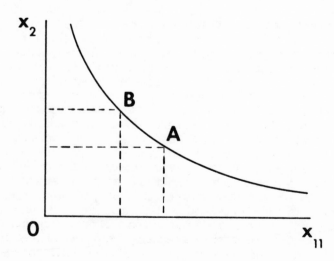

There are, therefore, more than one isoquant of interest in the analysis of annual crop production in mixed cropping systems. We may imagine a curve such as Y_1 in figure 4.3, which is the envelope of a large number of individual isoquants, each representing a constant quantity of output of a specific crop mixture, y_1, obtained by use of variable quantities of cropping labor, x_{11}, and land area, x_2, other things being equal. Among the "other things" we must assume that each y_1 is capable of being produced within a given range of soil and climatic conditions and subject to the condition that CEB not decrease. We will call this convenient envelope curve a "meta-isoquant," or potential isoquant.[8] Again, all points on Y_1 will be efficient points of production, since all are coincidentally points on individual isoquants.

Economic efficiency in using resources in production is indicated by points of tangency between isoquants and lines showing the relative prices of cropping labor and land area. If the farmer is not restricted to the choice of one crop mixture to produce, however, economic efficiency in using resources in production will be indicated by the points of tangency of price lines to the meta-isoquant over the range of the farmer's production possibilities, and not uniquely to one of the many individual isoquants.

By way of illustration let us make a greatly oversimplified input-output comparison of two of the upland crop mixtures studied by Norman and co-workers in northern Nigeria. By fitting production functions to their data on inputs and outputs it would be possible to derive isoquants for each of these crop mixtures. Let y_1 be one such isoquant, representing output of 1,000 lbs (a crude aggregate weight measure of output sufficient for our illustration) of millet/cowpeas/red sorrel mixture. If the price line between labor and land is indicated by P^0, economic efficiency will be attained at point A.

Now suppose the price relation between labor and land changes to P^1 such that land becomes more expensive relative to labor. Economic efficiency in production of millet/cowpeas/red sorrel will be maintained by shifting to point B within the production period, using more of the cheaper labor and less of the expensive land. However, it appears from the data summarized in table 4.1 that it takes more cropping labor input and less land to produce

Figure 4.3 Technical Change in Crop Production along Meta-Isoquant (Input Substitution, Production Function Fixed)

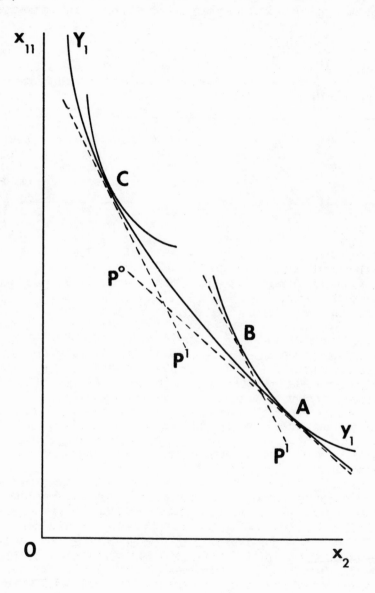

1,000 lbs of millet/sorghum/cowpea mixture than to produce 1,000 lbs of millet/cowpeas/red sorrel mixture, other things being equal. It will now be advantageous to move along the meta-isoquant to point C by switching within the production period to production of millet/sorghum/cowpeas.

TABLE 4.1
Land and Cropping Labor Output Measures, Upland Crops, Northern Nigeria

Crop mixture[a]	Average aggregate output (lbs/acre)	Cropping labor[b] (man-hrs/ acre)	Average output (lbs/ man-hr)
	(1)	(2)	(3)
Sorghum	582	210.21	2.77
Groundnuts	383	127.61	3.00
Millet/sorghum/ groundnuts/cowpeas	668	208.93	3.20
Millet/sorghum/ cowpeas/red sorrel	605	152.73	3.96
Millet	657	155.54	4.22
Millet/late millet/ sorghum/cowpeas	634	145.62	4.35
Millet/cowpeas	696	149.02	4.67
Millet/sorghum/cowpeas	856	181.66	4.70
Sorghum/deccan hemp	994	203.22	4.89
Millet/sorghum	962	193.10	4.98
Millet/cowpeas/red sorrel	434	70.41	6.16

[a]Data on millet/groundnuts/cowpeas and cassava were incomplete.
[b]This corresponds roughly to x_{11} input. The amounts shown may be understated, since the authors say (p. 90) that some ridging labor is included under "cultivating."
Source: D. W. Norman, J. C. Fine, A. D. Goddard, W. J. Kroeker, and D. H. Pryor, A Socio-Economic Survey of Three Villages in the Sokoto Close-Settled Zone, 3. Input-Output Study, Vol. 1, Text, Samaru Miscellaneous Paper 64, 1976. Column 1: table 65, p. 96; column 2: table B15, pp. 185-190; column 3: calculations by the author.

In this process, although input proportions have been changed further, there has still been no change in the parameters of the production function.

It is to be emphasized that the nature of low-resource agriculture in Africa makes such switches in price lines between sets of inputs likely in the very short term. Variations in rainfall patterns are common, for instance, and can affect the relative expensiveness or cheapness of resources like labor and land. For the farmer, substitution of one crop mixture for another on very short notice becomes one means of coping with this variability.

MARGINAL ANALYSIS OF PRODUCTIVITY

Were y_1 the only output of low-resource African agriculture, it would be a relatively simple task to determine, through application of neoclassical marginal analysis, the most technically and (knowing relative "price" ratios) economically efficient long-run potential combination of resources in the production of any one of the possible crop mixtures. The production function hypothesized in equations (1) and (2), however, is one in which annual crop output and CEB are joint products. This fact considerably complicates the determination of the efficient point of production.

The output of y_1 depends not only on a set of inputs in equation (1) that directly influence y_1 (let us call them x_i), but also on a set of inputs in equation (2) that indirectly influence y_1 through their influence on the output y_2 (let us call them x_j). We can see that the total effect of these inputs on y_1 depends in the long run on the summation of the marginal product of each multiplied by the change in quantity of each.

From equation (1):

$$dy_1 = \sum \frac{\partial y_1}{\partial x_i} dx_i + \sum \frac{\partial y_1}{\partial y_2} dy_2 \qquad (4)$$

From equation (2):

$$dy_2 = \sum \frac{@y_2}{@x_j} dx_j \qquad (5)$$

Substituting (5) in (4):

$$dy_1 = \sum \frac{@y_1}{@x_i} dx_i + \sum \frac{@y_1}{@y_2} \cdot \frac{@y_2}{@x_j} dx_j \qquad (6)$$

and simplifying, we obtain:

$$dy_1 = \sum \frac{@y_1}{@x_i} dx_i + \sum \frac{@y_1}{@x_j} dx_j \qquad (7)$$

Particular interest in our production function attaches to the input labor because of the widespread impression among agricultural economists that low average productivity of labor in Sub-Saharan African agriculture is an obstacle to the diffusion of technological innovations.[9]

Equation (7) shows that in terms of labor the total increase of annual crop output depends on the way in which labor influences crop output directly (x_i) and indirectly through CEB (x_j), as well as on the quantity of labor applied to annual cropping and conservation of equilibrium biomass. This brings us to the question of how best to achieve higher output within the constraints imposed by the production function posited. But before turning to an examination of how the whole production function can shift upwards, let us carry the analysis of marginal productivities of factors further.

We are interested in the first instance from a technical point of view in the marginal coefficients of production of x_{11} and x_{12}, labor used in annual crop production and labor used in conserving equilibrium biomass, respectively. These are derived by definition from the negative of the slope of the concave curve in figure 4.1, which represents a given level of inputs:

$$RPT_{12} = -\frac{dy_2}{dy_1} \qquad (8)$$

Taking the total differential of the equation of the product transformation curve for the general case of variable inputs,

$$x = w(y_1, y_2) \tag{9}$$

we have in the case of labor input, x_1:

$$dx = \frac{\partial x_1}{\partial y_1} dy_1 + \frac{\partial x_1}{\partial y_2} dy_2 = w_1 dy_1 + w_2 dy_2 = 0 \tag{10}$$

where w_1 and w_2 are the marginal coefficients of production of y_1 and y_2, respectively.

Technical complementarity between y_1 and y_2 requires that if one product is increased in quantity, the marginal coefficient of production of the common input used in the production of the other product decreases. That is, in equation (10) if $w_{12} < 0$, then dx_1/dy_1 decreases as y_2 is increased, which is to say that the marginal physical product of x_{11} increases.

This may be seen by substituting equation (3) in equation (2), making x_{11} an input common to both equations (1) and (2):

$$y_2 = f((x_1 - x_{11}), y_1; x_4, x_5) \tag{11}$$

It follows from our analysis of annual crop output and conservation of equilibrium biomass as joint products that it is not possible to make any statement about marginal or average productivities of one variable factor in the production of one of these products without taking the other product into consideration. To do so would be to assume that output consists of homogeneous units. As we have seen, in dealing with joint production we can make no such assumption.

In the choice of crop mixture (i.e. the composition of y_1), however, we have enough information to give us an approach to measuring such factor productivities. Output of y_2 depends heavily on the farmer's management practices, among which choice of crop mixture is one important dimension. Because crop mixtures entail widely varying combinations of input use, represented by movements along the meta-isoquant, we know that the productivities of variable inputs vary according to the composition of the output y_1. This is a strong conclusion when it is remembered that in African low-resource agriculture the composition of annual crop output represents in a very real sense one solution to the non-trivial problem of producing output without damaging the resource base. It will be seen that the measurement of input productivities in African low-resource agriculture requires practically a field-by-field investigation.

Our analysis of marginal physical products of variable inputs can be extended further with respect to labor input in its various forms. Equation (1) may be rewritten (taking equation (3) into account):

$$y_1 = f((x_1 - x_{12}), x_2, x_3; y_2) \tag{12}$$

Holding x_2 and x_3 constant and differentiating (12) with respect to x_1:

$$\frac{\partial y_1}{\partial x_1} = \frac{\partial y_1}{\partial x_{11}} + \frac{\partial y_1}{\partial y_2} \cdot \frac{\partial y_2}{\partial x_{12}} \tag{13}$$

So long as all terms on the right-hand side of equation (13) are of the same sign (positive), it follows that the rate of change of y_1/x_1 is greater than the rate of change of y_1/x_{11} in the relevant range.

Thus, we can conclude that in the long run, where CEB is variable, the total marginal physical product of labor is higher than the partial marginal physical product of labor in annual cropping, but declines at a steeper rate with additions of labor inputs, whatever the labor input (other than zero) into conservation of equilibrium biomass. This is a strong conclusion.

Note that this conclusion is valid even holding other factor productivities unchanged. It is not the same phenomenon, therefore, as diminishing returns to labor resulting from population pressure forcing cultivation of marginal land.[10]

It is likely that the "average" productivity of labor in crop production in Sub-Saharan Africa is higher than usually measured because labor investment in conservation works is often lumped with labor input in annual crop production. Confusion between labor employed to produce one output with labor employed to produce another accounts for the paradox that, despite the fact that labor has been identified as the scarce factor in African agriculture, the rate of adoption of labor-saving technology has been so low. It is now recognized that empirical calculations of average and marginal productivities of labor in African low-resource agriculture have resulted in great variability due to discrepancies of what is being measured as labor input.[11]

If the new technology fails to increase equilibrium biomass production, the marginal physical product of x_{12} falls rapidly, even with x_{11} constant or rising due to introduction of animal traction. In such instances, one presumes, the decrease in marginal physical product of total labor input is at a high enough rate to prevent adoption. It is not the technology itself that is inefficient or uneconomical. It is simply that the consequences for the equilibrium biomass of using the new technology are too costly in terms of labor.

Responding to population pressure the Sérèr have sent their large livestock away from their villages during the rainy season so as to be able to cultivate normally fallow land with a millet and groundnut rotation, or else have inserted a ring of continuously-cropped fields, which they fertilize with animal droppings, between the pombod and the large fields. These changes substitute labor for land. But the measures needed to maintain CEB under either of these "new" systems involve the farmers in additional labor input. If the labor input required by this "land-saving" technical change becomes onerous, farmers' labor is insufficiently compensated at the prevailing shadow price of labor (which applies to labor expended in CEB maintenance as well as in annual crop production), farmers will revert to their old system, making do with lower yields but preserving their resource base. It is only under compulsion that they would be expected to act otherwise.

Development projects that furnished farmers with animal traction equipment on the premise that labor being scarce, introducing a technological innovation that raised the productivity of labor would result in intensification, found that farmers used the equipment to extend the area cultivated, instead of concentrating their labors on the same or a smaller area but changing their cultivation practices to farm it more "intensively." Thus, it appears that even with a correct analysis of relative factor availabilities one can easily err in predicting the farmer's rational response to proposed changes in the low-resource farming system, unless one explicitly allows for the effects of changing input ratios or technology on the equilibrium biomass. Farmers cannot be expected to sacrifice CEB to gain higher productivities of labor and other inputs in annual crop production.

Since labor input in low-resource agriculture responds, through the mechanism of price changes, to changes in total marginal productivity of labor in our joint production function, the analysis of marginal productivity of labor in annual crop production alone is not sufficient to determine the economically efficient allocation of inputs. This is because changes in CEB shift the isoquants (and perforce the meta-isoquant) defined by equation (1) in a non-symmetrical way, as described more fully in the following chapter. The problem of pinpointing the final location in input space of the economically efficient utilization of cropping labor and other variable inputs is, therefore, more difficult than applying neo-classical marginal analysis to productivities of inputs in annual crop output.

NOTES

1. Earl O. Heady, Economics of Agricultural Production and Resource Use (New York: Prentice-Hall, Inc., 1952), p. 766.

2. For the conductor of farm surveys, mixed cropping poses problems that are by no means negligible. (See Arthur J. Dommen, "Producing Good Farm Surveys," Intech Paper No. 75-2 (revised May 1976), p. 38.) Some possible approaches to overcoming the difficulty for data collection and analysis presented by mixed cropping are described in James L. Stallings, Data Collection in Subsistence Farming Systems: A Handbook (Department of Agricultural Economics and Rural Sociology Paper No. 38, Auburn University, Alabama, October 1985), pp. 23-28.

3. Fresco, Cassava in Shifting Cultivation, pp. 133-134.

4. Keith Hart, The Political Economy of West African Agriculture (Cambridge: Cambridge University Press, 1982), p. 10.

5. Among the more interesting attempts to systematize farmers' huge annual variations in sown area in the SAT zone is Becker's, in which he hypothesizes an "implicit price" mechanism. (John A. Becker, "An Analysis and

Forecast of Cereals Availablity in the Sahelian Entente States of West Africa," final report to AID under contract AID/CM/afr-c-7320, 1974.)

6. As used, for example, by De Schlippe, Shifting Cultivation in Africa, p. 37.

7. As Carlson put it:

> When the proportions between the different products varies with different outputs, there is no longer a homogeneous output unit to which the productivities, costs, and revenues of the different services can be related. Nor is it possible to relate these magnitudes separately to the different products and to calculate their individual costs and revenues, since a change in one of the products will generally influence the technical, the cost and the demand relations of the others. This interrelationship between the different products is the characteristic feature of joint production.

(Sune Carlson, A Study on the Pure Theory of Production (Clifton, N.J.: Augustus M. Kelly, 1974 (repr.)), p. 76.)

8. This analysis owes much to Hayami and Ruttan, whose metaproduction function is defined as the envelope of commonly conceived neoclassical production functions, this locus being that of "the most efficient production points available." (Yujiro Hayami and Vernon W. Ruttan, Agricultural Development: An International Perspective (Baltimore: Johns Hopkins University Press, revised edition, 1985), pp. 134-135.) A major difference between the concepts of metaproduction function and meta-isoquant is that the former holds in the secular period of production, whereas the meta-isoquant is a function that accommodates crop mixes in the production period, making the task of data assembly for empirical verification less onerous.

9. Christopher L. Delgado and Chandrashekhar G. Ranade, "Technological Change and Agricultural Labor Use," chapter in Mellor, Delgado, and Blackie (eds.), Accelerating Food Production, p. 134. Delgado and Ranade maintain that in general the marginal product of labor in African agriculture is higher than that in Asia, whereas the average product of labor is lower.

10. For a view often expressed by economists, see the citation in Boserup, Conditions of Agricultural Growth, p. 31, fn. 2.

11. Delgado and Ranade, op. cit., p. 121.

5

Innovation

Having analyzed the influence of cropping labor input and labor input into conservation of equilibrium biomass in joint production, it now becomes possible to focus more sharply on the manner in which production change takes place in African low-resource agriculture and the conditions under which labor use may be expanded without loss of productivity.

Many empirical studies have turned up figures showing that cropping labor input is roughly similar in different cropping systems. The contradiction between this observation and the statement previously made that African crops have significantly different labor requirements is only an apparent one. Large amounts of labor for the production of crops like yams and rice go into conservation operations, like mounding, ridging, and fertilizing. These operations account for most of the variability in total labor input across crops.

Higher intensities of production imply a higher proportion of total labor going to soil-conserving works like dykes and mulching. In the SAT zone, such works are much more evident in the infields than in the outfields, where occasional animal droppings, rotations of legumes with grains, or planting acacia trees may be relied on by farmers to retain soil fertility and where crops may be sown without even a preliminary plowing.

In the three Nigerian villages studied by Lagemann, the tasks of ridging and mounding absorbed fully 16 percent of total labor input in field cropping in the high-density village, while they were negligible in the low-density village.[1] In the high-density village (where total labor input was also highest), ridging and mounding were ways in which the villagers concentrated their intensive efforts at preventing soil fertility from falling in the absence of fallowing. These data did not even include labor time spent on tree crops, which were important in the region and particularly so in the high-density village, where Lagemann used the expression "multi-storey cropping system" to describe the way in which tree crops were integrated with field crops.[2]

EFFICIENT CROP MANAGEMENT SYSTEMS

In the crops of the SAT zone village in Burkina Faso studied in detail by Prudencio, cropping labor per hectare, once the labor involved in conservation operations had been deducted, was rather noticeably similar across the different rings, or crop management systems, that his observations allowed him to distinguish (table 5.1, col. 3). He states, moreover, that the proportions in column 2 would likely be lower for rings 1 and 2 because they have not taken into account all the manure transportation informally carried out during the dry season, these two rings receiving by far the heaviest doses of manure.

Prudencio's measurements of output in value terms, on the other hand, showed striking differences across cropping rings. These differences are due to the variable composition of crop output, as already mentioned. Moreover, the value of output varies in a systematic manner from one ring to another as one moves progressively from the compounds to the outfields (table 5.1, col. 4).

These data enabled Prudencio to calculate separate production functions (table 5.2) for each of the rings, or crop management systems, that he observed, as we have seen.[3]

Prudencio's rings afford an instantaneous picture of a set of crop management systems. Each is efficient. Indeed, one may argue plausibly that if the cultivating methods used in each one of these rings were not efficient, the farmers would have abandoned them long ago.

African farmers have sustained such contiguous sets of crop management practices over many generations. The farming systems of the West African SAT are particularly illustrative of such an equilibrium of systems, since their cropped fields are laid out, Von Thuenen-style, in concentric rings around the compounds of the farmers (see figure 5.1, drawn after Pélissier's diagram of the Sérèr village of Diohione in the Sérèr country of Senegal).

To sustain such differences in the composition of crop output over long periods of time (generations), however, the management systems of the more "intensive" rings must maintain soil fertility at a higher level than that of the less "intensive" rings. That soil fertility differences form the basis for the factor productivity differences shown in table 5.2 was demonstrated by Prudencio.

Time is graphed along the horizontal axis and area along the vertical axis in figure 5.2, which illustrates his method. The vertical dimension of the diagram need not be bounded strictly, since it is likely that over time more or less land comes under the control of the farming community. However, for simplicity of illustration, a horizontal upper bound is adopted. Also, I will refer here, like Prudencio, to the simpler concept of soil fertility rather than to CEB.

TABLE 5.1
Labor Input, Intensity of Production, and Value of Output, Nonghin Village, 1981

Cropping ring[a]	Average labor input (hrs/ha)	Percent of labor in cropping (percent)	Average cropping labor (hrs/ha)	Average value of output (CFA francs/ha)
	(1)	(2)	(3)	(4)
1	619	81	501	169,237
2	697	88	613	69,278
3	442	88	389	46,130
4	522	93	485	41,605
5	514	95	488	24,373

[a]Cropping rings are at greater distances from compound according to increasing numerical order.
Sources: Column 1: Prudencio, op. cit., table 11, p. 82; column 2: ibid., p. 106; column 3: calculated by the author; column 4: ibid., table 11, p. 82.

TABLE 5.2
Factor Productivities, Nonghin Village, 1981

Factor productivity measures	Cropping rings[a]					
	5	4 and 5	All	2, 3, and 4	2 and 3	1
Cultivation frequency[b]	30	45	50	80	94	96
Kilograms of combined output						
Average physical products:						
Land area (A)	388	489	761	1,101	1,512	2,109
Marginal physical products:						
Land area (C)	142.6	91.4	383.3	484.8	805.2	709.5
Hand-tool labor, seedbed preparation, planting (C)	-1.72	-6.88	-0.59	1.80	1.89	-1.91
Hand tool labor, weeding (C)	0.44	1.06	1.39	1.90	1.09	1.44
CFA francs						
Average value products:						
Land area (A)	24,025	27,635	39,595	54,237	60,483	200,617
Marginal value products:						
Land area (B)	8,156	2,656	19,539	17,900	32,725	68,182
Hand-tool labor, seedbed preparation, planting (B)	-1.5	-56.0	-41.8	62.5	76.7	-206.0
Hand-tool labor, weeding (B)	28.6	85.5	107.7	127.8	39.5	149.0

Note: Productivities are calculated at the mean, per hectare for land and per hour for labor.
[a]Cropping rings are at greater distances from compound according to increasing numerical order.
[b]See p. 30, above.
Source: Prudencio, op. cit.: (A) table 27; (B) table 30; (C) table 34. Reprinted by permission.

Figure 5.1 Concentric Ring Farming System

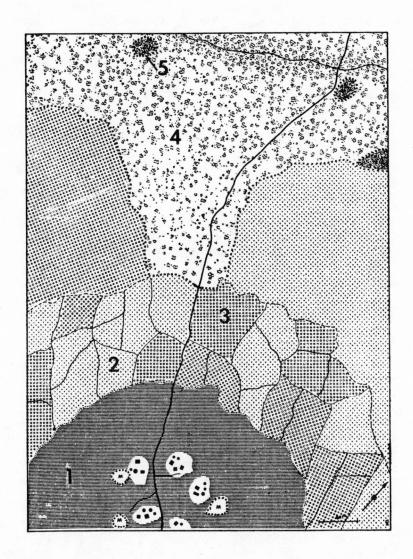

Legend: 1 Pombod. 2 Groundnuts. 3 Match. 4 Fallow and bush. 5 Sorghum.

Source: Pélissier, <u>Les Paysans du Sénégal</u>, figure 23, p. 295. Reprinted by permission.

Let the whole area of the community's land be fallowed in year 0. In years 1, 2, and 3, the land becomes occupied by farmers A, B, and C, successively, as shown. However, a portion of the community's land remains under fallow, with natural soil fertility represented by q_o^o.

Assuming a uniform system for managing soil fertility, the change in soil fertility for occupant A between years 0 and 3 is $\Delta = (q_o^o - q_3^a)$. The cross-section approach estimates that change to be $\widehat{\Delta} = (q_3^o - q_3^a)$. However, q_3^o is a better reference than q_o^o since it indicates what the fertility of the land occupied by farmer A would have been in year 3 if it were not being cultivated. This accounts for the effects of non-management variables such as weather.

The cross-section approach therefore estimates the dynamic changes in soil fertility, $\widehat{\Delta}$, by comparing the three soil fertility gradients (1) q_3^o with q_1^c, q_2^b, and q_3^a; (2) q_1^c with q_2^b and q_3^a; and (3) q_2^b with q_3^a.[4] Prudencio's investigation showed that such soil fertility gradients exist.

The tendency for African agricultural systems to exhibit several different degrees of intensity of production simultaneously leads to a logical inference from cross-sectional observation to variation through time. Prudencio's ring observations make possible a dynamic translation of a static picture of contiguous crop management systems to a process of moving from less "intensive" to more "intensive" systems.

In terms of economic theory, when African farmers recombine the same crops and the same tools in new ways so as to obtain their desired output they are shifting their production functions upward. They vary their resource management techniques in the absence of any technological change. Prudencio, to my knowledge, was the first agricultural economist to formulate this habit of African farmers in terms of an upward shift of the production function.[5] His attention to the significance of such a shift was attracted by his observation that the pattern of crop substitution seemed to be not of low-fertility-demanding crops for high-fertility-demanding crops, as might have been expected, but vice versa. Prudencio emphasizes the gradual nature of this shift in production function.

The proven differences across Prudencio's rings, and the transition from one management system to another implied by these rings is entirely attributable to differences in the ways in which crops and animals are managed. This, it would appear, is the kind of difference

Figure 5.2 Soil Fertility Gradients

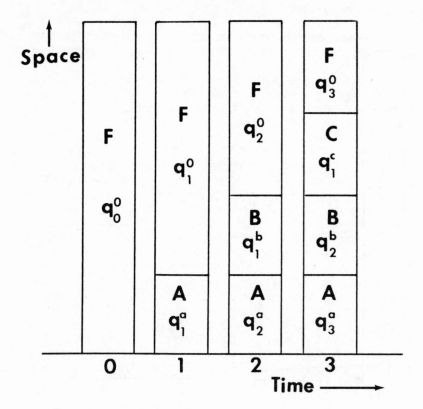

96

that Schumpeter had in mind when he defined innovation: it illustrates not technological innovation, but what Schumpeter called organizational innovation:

> We will now define innovation more rigorously by means of the production function. . . . This function describes the way in which quantity of products varies if quantity of factors vary. If, instead of quantities of factors we vary the form of the function, we have an innovation.

And again:

> . . . we will simply define innovation as the setting up of a new production function. This covers the case of a new commodity as well as those of a new form of organization or a merger, or the opening up of new markets . . .[6]

If we were to try to graph the rather systematic variation in factor productivities across cropping rings we would get a graph like figure 5.3. Here, <u>at the same level of input (variable capital) use,</u> total production is higher, and consequently average and marginal productivities are higher, in ring 1 than in ring 5.

Figure 5.3 Prudencio's Rings

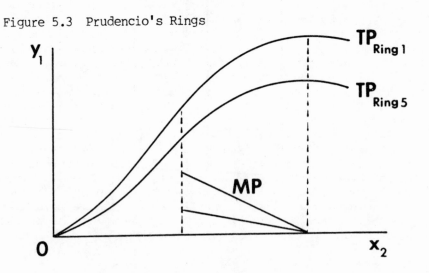

The difference between the total product curves for ring 5 and for ring 1 in figure 5.3 also illustrates, by logical inference, change in the time dimension following farmers' change from one crop mixture to another. In the short run, only y_1 is variable in our production function, with y_2 being a fixed factor in equation (1). Still in the short run, the shift corresponds to a movement along the meta-isoquant in input space. Recalling figure 4.3, as input proportions change, use of a variable input can expand with no shift in the production function.

But production of a crop mixture demanding a higher level of soil fertility can be sustained only if CEB is increased. This requires labor investment in conservation works like bunds and ridges. In terms of our equation (2), y_2 output rises. In turn, y_2 in equation (1), which is a parameter in the short run, increases, increasing output of y_1.

In other words, both y_1 and y_2 outputs are now increasing. The situation is shown in figure 5.4, which this time shows total labor (x_1) along the horizontal axis and total outputs (y_1 and y_2) along the vertical (assuming additive units of measurement of crop output and CEB output). Total labor input can (gradually) expand considerably without loss in labor productivity (MPP). The increased labor "invested" in CEB boosts output greatly. We may call this CEB-augmenting organizational innovation, since organizational innovation, which is labor-intensive, ends up increasing the "supply" of CEB.

The process just described is truly organizational innovation because the parameters of the production function have changed. It is no longer a case of technical change involving input substitution. The original change of crop mixture has brought about a new way of organizing resources for production. In a system dependent on fallowing, since there is a tradeoff between level of edaphic fertility and cycle length, the resulting organizational innovation may assume the form of a shortening of cycle length. Indeed, such a shortening of cycle length is sometimes taken to be the definition of intensification. On the other hand, a decrease in labor input into CEB with little or no change in cycle length will result in a decrease in CEB, as appears to have been the case with the newly opened savanna fields on the plateau outside the valleys in the Kwango-Kwilu.[7]

EFFICIENT INTENSIFICATION

Up to now, we have used the term intensification in a rather loose manner to describe the apparent ability of African farming communities to adjust the use of the resources at their disposal so as to sustain relatively high levels of output. This adjustment involves a process corresponding to Schumpeterian organizational innovation. It is now time to define intensification more rigorously in terms of our production function.

The central operative concept of efficient intensification in agricultural production, regardless of the type of innovation involved, is the expansion of variable input use over the long run without serious loss of factor productivity. In the English agricultural revolution, land productivity and to a lesser extent labor productivity increased as a result of the changes brought about by the Norfolk rotation. Labor use expanded greatly. In terms of neoclassical production theory, the elasticity of output with respect to labor expanded.

We have seen in figure 5.3 how organizational innovation in African agriculture, rather similar in nature to that in the English agricultural revolution in that it consists of rearranging the way productive resources are used, can raise the annual crop output and at the same time intensity of cropping labor use. In the longer term, necessary increases of labor input into CEB (preparing fallow, terracing, bunding, mounding, and similar conservation works), other things being equal, will further shift the total product curve gradually upward and to the right, as in figure 5.4. The marginal physical product of a given amount of cropping labor will automatically rise in this case. Labor use can expand without loss of productivity. Further, as CEB increases we would expect marginal physical products of the other variable factors in equation (1) (i.e. land area and variable capital) to themselves increase. Conversely, if CEB should decrease, the opposite from the preceding will occur and the marginal physical product of cropping labor (and of other variable inputs) decrease.

Over time, we would expect the labor investment necessary to achieve a measure of higher output to vary widely by location because, while annual crop output in the preceding year exerts at all times a depressing influence that must be overcome, the total elasticity of y_2 depends heavily on the level of x_4, intrinsic

Figure 5.4   CEB-Augmenting Organizational Innovation

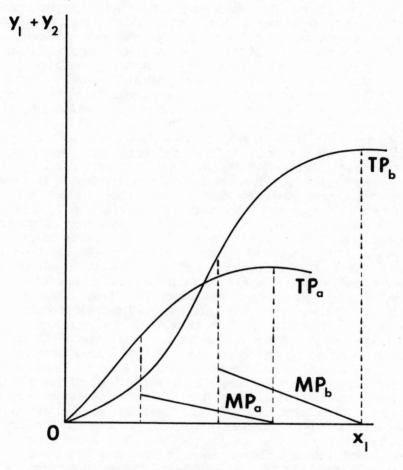

a   Before   innovation

b   After   innovation

fertility of land. If the initial level of soil fertility is low, the range of suitable crop mixtures will be restricted, reducing mobility of factors like labor and land area, and making intensification difficult; but it is nonetheless possible.

The change in the composition of output that accompanies the gradual shift upwards of the production function due to "investment" in CEB can best be seen in input space in figure 5.5. Let us graph variable inputs land area (x_2) and capital (x_3) along the vertical and horizontal axes this time, holding other factors constant. Y_1' represents the unit meta-isoquant before intensification. Let us say the farm is producing millet and white sorghum at point A. The farm is able to "save" land by replacing millet and white sorghum by red sorghum, moving along the meta-isoquant to point B by means of a process of technical change as described previously.

Since red sorghum is more fertility-demanding than the millet and white sorghum mixture, this change in crop mixture implies the necessity of maintaining a higher level of soil fertility than before by more intensive application of manure and other fertility-restoring inputs. The farmer must "invest" labor in CEB. A gradual increase in CEB shifts the unit meta-isoquant to a new position, Y_1''. The new point of production is point C.

In sum, the change implied by efficient intensification is the result of two inter-related moves: from point A to point B initially, then from point B to point C. The former results from an adjustment of the x_i inputs (corresponding to the first term on the right-hand side of equation (7)). The latter results from a homothetic movement of the meta-isoquant occasioned by the indirect effect of the x_j inputs (corresponding to the second term on the right-hand side of equation (7)). The same analysis can be applied with respect to other combinations of inputs.

Once the additional labor investment in CEB output has been made, higher efficiency of annual crop output is achieved relatively easily because the "direct" inputs work better (more efficiently) than they did before thanks to the production possibilities opened up by higher CEB and crop mixture changes. The exact amount of output added per unit of the "direct" input in this process of intensification depends, of course, on the partial elasticity of output with respect to that input. But there is a strong probability that in low-resource agriculture such

Figure 5.5 Efficient Intensification: Land–Saving Technical Change Followed by Meta–Isoquant Shift (Due to Production Function Shift)

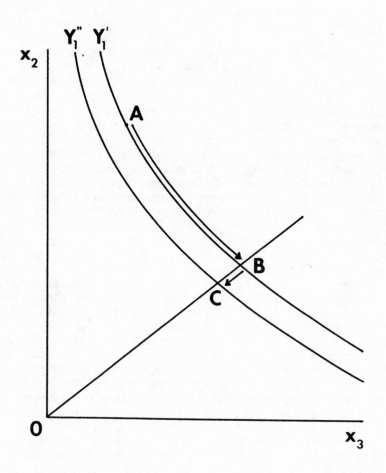

partial elasticities are in most instances positive. The important point is that as long as this kind of change is possible there is no reason to assume that low-resource agriculture is producing at its peak capacity and is not susceptible of further improvement.[8]

Mixed Cropping as a "Disguised" Variable

Because growing crops in mixtures not only determines the sustainable level of soil fertility in the long run, but also facilitates technical change in the short run and accelerates efficient intensification of production in the long run, the practice of mixed cropping is enormously important in Sub-Saharan African low-resource agriculture. Crop mixture emerges as a "disguised" variable whose role in determining resource allocations both in the short run and long run has been much underestimated.

Change of the kind described here is gradual, and therefore is often not recognized for fundamentally altering the parameters of the production function. Yet, in this theoretical analysis efficient intensification has the same effect as the much more dramatic change implied by the introduction of high-yielding crop varieties, as occurred in the Green Revolution in Asia.

Both types of change are examples of Schumpeterian innovation in that they involve altering the parameters of the production function. While the Green Revolution type of change can be safely subcategorized as technological innovation, involving as it does new "land-augmenting" technology, the case of efficient intensification in African agriculture can be subcategorized as organizational innovation, because its essential element is a new way of organizing resources for production and conservation of soil fertility.

Organizational innovation may indeed lead to technological innovation in this process. The additional labor investment in CEB output raises the total labor requirement significantly, as shown in figure 5.4. Here, labor use can expand to the right considerably without suffering lower marginal product than was the case before innovation.

The higher productivity of this additional labor may lead to substitution of one farm tool for another or of a high-yielding variety for a traditional variety. Tools used in land preparation, made necessary by more intensive crop mixtures, offer an example. Turning over the soil results in a beneficial bringing to the surface of soil nutrients. Performing this task with a hand hoe, however, may be arduous, whereas a donkey-drawn plow may be well suited to the task and now becomes profitable.

Substitution among relevant inputs will continue as a result of increasing CEB until returns to factors conform to economic returns and economic efficiency is restored. Technological change may lead to even greater efficiency. Application of additional fertilizer, for example, may stimulate weed growth so that substitution of an animal-drawn cultivator for the hand hoe becomes even more advantageous.

Unlike the cases previously cited, animal traction here assumes a rationale of its own from the farmer's point of view other than merely expanding cultivated area. Here, it becomes a means of "saving" labor in a labor-intensive land preparation operation and in one of the most labor-intensive cropping operations. The important point is that the determining factor in adoption of animal traction is economic.[9] But the technical conditions for innovation must exist before the economic forces determining choice of technique can impinge on relative factor scarcities.[10] This finding offers a theoretically satisfactory solution to the "puzzle" noted by Pingali, Bigot, and Binswanger to the effect that animal traction is only adopted at higher population densities.[11] Lower population densities mean that, although agricultural labor may be scarce, the incentive to organizational innovation that paves the way for labor-augmenting technological change may be lacking.

Characteristic Features

The characteristic features of efficient intensification as applied to African low-resource agriculture may now be summarized. In the case of figure 5.4, first, total labor input expands without loss of labor productivity. Second, this expansion of labor use is accompanied by an increase in outputs both of annual crops and CEB. The outward manifestations of these changes are a change in crop mixture, a higher proportion of labor invested in soil-conserving works and cultivating practices, and possibly a shortening of cycle length. Technological change may also be associated with the process.

We may now rigorously define efficient intensification of production in African low-resource agriculture as the process of attaining more than proportionately higher factor productivities of variable inputs through gradually releasing the constraint on sustainable expanded use of these inputs imposed by conservation of equilibrium biomass.

Generalizing the case of total labor input (illustrated by figure 5.4) to all inputs, we can state the above definition in terms of the production function.

First, comparing input use before (a) and after (b) innovation, we have:

$$\sum (x_i + x_j)_b > \sum (x_i + x_j)_a$$

where, it will be recalled, x_i are inputs affecting y_1 output directly and x_j are inputs affecting y_1 output indirectly.

Or: $\sum (x_i + x_j)_b = (1 + \alpha) \sum (x_i + x_j)_a$ where $\alpha > 0$ (14)

Furthermore, from the proposition familiar from the calculus that the variation of a function resulting from the simultaneous variation of all arguments equals the sum of the variations in the function ascribable to independent variations of the arguments, we can write the relation of output before to output after innovation as follows (for the region of figure 5.4 where $TP_b > TP_a$):

$$\sum (x_i f_i)_b + \sum (x_j f_j)_b > \sum (x_i f_i)_a + \sum (x_j f_j)_a$$

where f_i represents marginal product of input i (and similarly f_j). Or:

$$\sum (x_i f_i)_b + \sum (x_j f_j)_b = (1 + \lambda) \left[\sum (x_i f_i)_a + \sum (x_j f_j)_a \right]$$ (15)

where $\lambda > 0$

Equation (14), which we may call the input use expansion condition, and equation (15), which we may call the output expansion condition, are together sufficient conditions for intensification. For intensification to be efficient, however, a third condition, which we may call the efficiency condition, is necessary:

$$\lambda > \alpha$$ (16)

Efficient intensification is a long-term condition, because marginal physical products are measured in such a way as to include CEB, which is a long-term product. Further, we can compare directly how the proportion of outputs y_1 and y_2 changes as a result of intensification.

From equations (5) and (7), we have:

$$\frac{dy_1}{dy_2} = \frac{\sum \frac{\partial y_1}{\partial x_i} dx_i + \sum \frac{\partial y_1}{\partial x_j} dx_j}{\sum \frac{\partial y_2}{\partial x_j} dx_j} \tag{17}$$

The right-hand expression in (17) may be either positive or negative. In the extreme case, the marginal product of x_5, which is always negative, can outweigh the sum of the positive marginal products of the variables x_{12} and x_4, making the denominator of (17) negative. This would make the ratio of y_1 to y_2 decrease as a result of intensification.

The normal case, however, would be the opposite, with the proportion y_1 to y_2 increasing as a result of intensification. The key role played by x_{12} should be evident from the fact that it influences both the denominator and the second right-hand term in equation (17). If the quantity of input x_{12} is insufficient, the expression dy_1/dy_2 decreases no matter how much input x_{11} is used.

In other words, in cases where the cultivation frequency increases as part of the intensification process, labor "investment" in CEB must more than offset the negative effect of higher cultivation frequency on CEB. An increase in cropping labor alone may compensate for some time for a loss of CEB due to lack of such "investment" labor, for changes in CEB are long-term in nature. But eventually this effort will fail. This appears to be the case of cassava cropping in the Kwango-Kwilu, for cassava is a low-cropping-labor-input crop substituting there for higher-cropping-labor-input crops.[12] Due to the difficulty of separating types of labor almost everywhere in African agriculture, the determination of how intensification affects total labor productivity must be a matter for empirical investigation.[13]

In figure 5.4, the two production functions cross. This results in a situation where at very low levels of cropping labor input the total output associated with the crop mixture grown on the higher-level-CEB land is actually less than that associated with the crop mixture grown on the lower-level-CEB land at the same level of total labor input.

In the village studied by Prudencio, cowpeas were a frequent component of crop mixtures. However, cowpeas grown in mixture with red sorghum on infields were invariably damaged by goats and chickens searching for forage and feed, both early in the growing season when the young cowpea plants were exposed to these animals, and late in the season when the red sorghum had been harvested, thus no longer providing protective cover for the later-maturing cowpeas. Only when labor was available in sufficient quantities to guard the crops did the cowpea/red sorghum mixture show its true output potential on the high-CEB fields.[14] Watching growing crops to chase away grazing animals, birds, and other predators is a normal labor requirement of low-resource agriculture, usually performed by children. To the right of their intersection in figure 5.4, the cowpea/red sorghum mixture outperforms the cowpea/white sorghum or cowpea/millet mixtures grown on the lower-CEB outfields, other factors being the same.

It will be apparent that the process of efficient intensification of production is not solely determined by the cultivation frequency. In the production function in equations (1) and (2), efficient intensification cannot be measured by any single ratio of variables, since outputs in one equation act as inputs in the other. The conditions set forth in (14) to (17) above are much more general in nature, and they represent an improvement in the analysis of the process of intensification.

In past attempts to analyze factor productivities in African low-resource agriculture, land is often treated as a fixed factor implicitly, if not explicitly. Intensification as Pingali, Bigot and Binswanger use the term, for instance, means "the movement from forest-fallow and bush-fallow systems of cultivation to annual and multi-crop systems, whereby plots of land are cultivated one or more

times a year."[15] Land is used more frequently, and
therefore additional labor is invested in the same permanent piece of land. But this introduces an unnecessary
difficulty into the analysis. A theoretical framework
which treats the specific surface aspect of land as a
variable factor, however, avoids the difficulty by making
the characteristic aspect of the intensification process
not the shift in the labor-land area ratio. Indeed, the
labor-land area ratio may remain constant as the process
occurs, or even decrease, since second- and third-year use
of the same previously fallowed land involves less of the
heavy labor of clearing necessary in the first year of
cultivation.[16]

VERIFICATION

Investigators of African low-resource agriculture
like Fresco (Zaire), Guyer (Cameroon), and Prudencio
(Burkina Faso) who have recorded changes in crop yields,
labor inputs, and other variables on the same land over a
period of years, usually by talking to farmers involved,
have produced insights which have greatly advanced our
knowledge of low-resource agricultural systems and their
sustainability. Only tentative conclusions, however, can
be drawn from the mass of empirical evidence gathered by
these investigators. A compilation of insights gleaned
here and there by investigators who approached African
agriculture with many diverse mental frameworks is no
substitute for systematic, rigorous hypothesis testing.
The start of such a test would be the collection of
data on inputs and outputs from one or more African
farming systems designed to measure significant relationships of crop production, soil fertility, and system
sustainability. Unfortunately, empirical verification of
the production function hypothesized in this book imposes
enormous data requirements, with respect to both inputs
and outputs, to enable calculation of parameter values
with some degree of confidence. The measurement of CEB
output alone, for instance, would require many years'
running of laboratory soil analyses providing the
observation of trends of conservation of equilibrium
biomass.
Although Prudencio's single-equation production
function is quite different from the one presented in this
book and includes a different set of explanatory

variables, his findings nevertheless suggest that the tendency for productivities of factors like land area and labor to rise with greater intensification is not implausible. Greater intensification in his investigation is depicted by the sequence of management rings going from the outfields towards the infields. Put in different terms, such intensification allows the elasticities of output with respect to variable factors like cropping labor to expand.

The results reported in this chapter do not by themselves constitute verification of the production function hypothesized in the previous chapter, since they stem from only a single investigation. However, they do demonstrate the feasibility of empirically verifying such a function by carrying out in-field data collection and analysis. In this sense, they have great scientific value.

NOTES

1. Lagemann, _Traditional African Farming Systems_, appendix tables 28 and 30, pp. 222 and 224. The total annual man-hours for the low-density village is misprinted and should read 15,213.5.

2. _Ibid._, p. 31.

3. See pp. 41-42, above.

4. Prudencio, "A Village Study," pp. 135-136.

5. _Ibid._, p. 320.

6. Joseph A. Schumpeter, _Business Cycles_ (New York: McGraw-Hill, 1939), Vol. I, pp. 87-88.

7. Fresco, _Cassava in Shifting Cultivation_, pp. 130-131. Shortening of fallows reportedly occurred "most dramatically" in the central valleys, not on the plateau.

8. The author of the survey cited in table 2.6 worked out an intensification method for increasing the total volume of output of the community he studied by a phenomenal 56 percent merely by reorganizing the existing use of resources, without introduction of new technology, at the same time preserving the resource base from damage. Hecq, "Le Système de Culture des Bashi," pp. 994-996.

9. There have been many myths said to account for the failure of adoption of animal traction in Sub-Saharan Africa--most of them non-economic in nature. Pingali, Bigot, and Binswanger, _Agricultural Mechanization_ gives a good rundown in Chapter 6.

10. De Wilde, discussing the failure of efforts to intensify (by which he means "to increase yields per unit of area by additional 'inputs,' including more labor") agricultural production in Africa, notes: "In most of these cases there was a failure to understand the relative scarcity of labor and land or to grasp the importance of maximizing returns to the scarcer factor." (John C. de Wilde, Experiences with Agricultural Development in Tropical Africa (Baltimore: The Johns Hopkins Press, 1967), Vol. I, p. 71.) In view of the above production function analysis, we can see that the failure of understanding resulted from the underestimation of labor inputs (to produce annual crops and maintain CEB), rather than from an error in the application of economic logic to the problem. This is reassuring to an economist.

11. Pingali, Bigot, and Binswanger, Agricultural Mechanization, p. 6.

12. Fresco, Cassava in Shifting Cultivation, p. 135.

13. Pingali, Bigot, and Binswanger, Agricultural Mechanization, p. 108, similarly, leaves this matter of total labor productivity a question open to further empirical investigation:

> Since our data did not include the overhead labor required for the training and maintenance of draft animals and for the land investments, such as destumping and leveling, associated with intensification, we are not in a position to decide whether intensification, including the associated shift in tools, leads to an increase or a decrease in the overall productivity of labor where the overall labor inputs include overhead labor.

It should be remembered that these authors define intensification in terms of frequency of cultivation per unit land area, meaning that they do not explicitly allow for changes in crop mixtures and the changes in quantity of output (in weight, value, or other terms) associated with such changes in measuring labor productivity.

The discussion of animal traction benefits in Chapter 7 of this source almost totally lacks any attempt to distinguish between animal traction for plowing (i.e. land preparation) and animal traction for weeding (i.e. crop cultivation proper). Yet, this is a most important distinction in the analysis of African agriculture. I am grateful to William Jaeger for this insight.

14. Prudencio, "A Village Study," pp. 90-91.

15. Pingali, Bigot, and Binswanger, Agricultural Mechanization, p. 4.

16. Binswanger has recognized the fallacy of measuring the economic benefits of innovations in African agriculture by the rate of increase in returns to land as a fixed factor of production. He concluded that this measure "can lead to highly misleading assessments" of such benefits. (Binswanger, "Evaluating Research," p. 2.)

PART THREE

The Population Question

6

Two Relations Between
Production and Population

The finding that the productivities of factors in low-resource agricultural production change not only as the intensity of production changes but also as cultivation practices are modified in accordance with locally available resources has obvious implications for the ability of low-resource agriculture to support denser populations. The production function posited in the previous section will now be used to explain the extraordinary resilience of African low-resource agriculture in the face of growing population pressure: its ability to accommodate a wide range of population densities, and its tendency to diversify agricultural production with increasing population density.

Certainly, we need to know the implications of the preceding analysis for the population support capacity of low-resource agriculture. But caution is always advisable in going from analysis of purely technical production relations to statistical inference of producer behavior once data availability and collection difficulties have been overcome. This caution applies with special force to Africa, where the village or the extended family, rather than the individual, is the basic decision-maker.[1]

First, there is the purely statistical problem of estimating the production function. Basically, the problem arises from the fact that biased and inconsistent OLS estimates of parameters may result from regression analysis when the observed input levels are partly dependent on behavioral influences subsumed in implicit derived demand functions for inputs. This problem is susceptible to solution by using food stocks, household food consumption, wealth, distance to fields, and other exogenous variables as instruments in the regression analysis. This, of course, compounds the data collection problem manyfold. The problem can otherwise be solved by assuming that input use depends on anticipated output rather than actual output, thereby eliminating the transmission of bias from the technical function (estimated) to the behavioral function (not estimated).

The second problem is even more serious. It is the well-known problem of "evolutionary determinism" against which I cautioned in an earlier footnote and concerns attribution of causality in use of cross-section data. In sum, by saying that organic manure applications per hectare of cropland in northern Nigeria show a positive correlation with population density (with both variables measured at a single point in time, as is the case with the data in table 3.1), I am not saying that there exists a necessary evolution to which rural populations in Africa are subject. In other words, the fact that changes in intensity of production and population density have been observed to take place through time, as well as to vary in the spatial dimension, does not necessarily mean that growing population density is the cause of intensification, or vice versa, although there may be a strong supposition to that effect.[2]

There is, however, no reason to believe that low-resource farming communities in Africa are not subject to the effects of supply and demand for agricultural products, in whatever units such supply and demand functions may be conveniently measured. Here, population growth acts as a strong demand shifter.

Population growth is, of course, not the only such shifter. Intensification of agricultural production may be favored by economic changes such as shifts in the demand for foodgrains on the part of nearby urban populations which make the marketing of surplus production more profitable for farmers. It may be favored by an increased need for cash income on the part of rural producers. Or it may be favored by social change in the rural sector of African countries, such as the breakdown of households into smaller units. But, by and large, increasing population density is the factor most visibly associated with the intensification of agricultural production.

In terms of the hypothesized production function, the direct effect of any of the above phenomena, ceteris paribus, is to reduce the second term in the denominator of the measure of cultivation frequency on page 32. This raises the value assumed by x_5 in equation (2), in turn immediately reducing y_2 output. The rural community, therefore, must take countermeasures to prevent a reduction of y_1 output in the longer term. Such countermeasures take the form of labor "investment" in CEB, thereby preventing the immediate fall in y_2 output and the longer-term decline in annual crop output that results therefrom. If such investment is sustained, it leads to an increase in CEB and efficient intensification of production. If not, the intensification of production that occurs is inefficient. An example will be given below.

THE INCONGRUENCE OF RESOURCES AND POPULATION DENSITY

Low-resource agriculture in Sub-Saharan Africa supports a wide range of population densities. A study of Cameroon shows average rural population densities by province varying from 3.7 to 82.0 persons per square kilometer, with densities in some regions going as high as 200 per square kilometer.[3]

Population densities in Sub-Saharan Africa are frequently not congruent with the distribution of physical natural resources. The tendency of the well-watered, more fertile southwest of Burkina Faso to remain relatively underpopulated, while the relatively arid Mossi plateau in the center of the country with its poor soils remains overpopulated has long puzzled observers. The dense farming populations of the Great Lakes Highlands of Rwanda, Burundi, and Kivu Province of Zaire go on for generation after generation producing a great variety of crops, perversely, it might seem, from soils that have major problems of acidification and alumuminum toxicity and lie on steeply sloping hills. Their agricultural system shows no less stability than that of the Sérèr in the "underpopulated" groundnut basin of Senegal and seems to be as durable.

Scientific investigation of the relationship between cultivation practices and population density in Africa originated in the colonial period when administrators saw the overpopulation of some areas and the degradation of land as problems linked by cause and effect. These investigators established the idea that the system of land use, rather than natural soil fertility, was the determining factor behind the amount of land required to support a given population.

Early astute observers of African agriculture like Allan and De Schlippe also noted the preference low-resource farmers had for cultivating lighter soils over heavier, more fertile soils in situations of relative land abundance. The argument often heard was one of work-saving: if one could cultivate three times as much poor land by furnishing the same effort as cultivating the heavier soils and get only half the yield, there was still a net gain.

Allan coined the term "critical population density" to relate the indigenous agricultural production system to population density on the basis of change "without damage to the land."[4] But he failed to take his inquiry one additional step further: instead of being a static warning signal of imminent soil exhaustion in an area considered by agricultural "improvers" for farming or resettlement, the critical population density was itself subject to increase over time as a consequence of changes in the production system.

In seeking the relationship between population density and natural resource distribution, African crops are a misleading indicator. The geographic distribution of crops is governed to a very large extent by rainfall amounts and patterns. The distribution of African agricultural tools, on the other hand, offers a more accurate indicator of population support capacity of different farming regions, since tools are man-made and their distribution is controlled by the societies that possess them and use them to produce crops.

As Raynaut has pointed out, there exists a distinct lack of concordance between distribution of types of tools regionally and the relatively advantageous or disadvantageous situation of these regions for agricultural production.[5] The short-handled hoe requires many more hours of labor to weed a given area than the long-handled hoe with a horizontal blade. But the former is used for weeding infields, rooting up weeds and burying them, while the latter is used for weeding outfields, skimming large areas relatively superficially. Labor is applied differently in the two cases. Moreover, sharp contrasts in tool inventories between regions appear to be lacking (other than ones of purely cultural origin). Instead, it seems that African villages have a large inventory of tools adapted to specific uses in agricultural production whatever region they happen to be in: clearing bush, felling trees, raking, hoeing, holing, weeding, and so forth. Thus, as Raynaut and others describe, there is today a tendency to extensive production even in regions where natural conditions such as rainfall and soil fertility levels would suggest specialization in intensive production, population density permitting.

The system of progressive intensification of agricultural production makes possible the high degree of self-sufficiency that is the hallmark of African village economy. Since the individual household farms plots of land scattered in all cultivation rings from infields to outfields, some of which are intensively farmed while others are extensively farmed, the crop output of each household comprises a wide array of crops. This diversity of output is not inconsistent with the needs of farmers who do not have profit maximization as their dominant objective function.

There is no reason why the intensification of production that accompanies increased population density in African farming systems need damage, much less destroy, the ecosystems in which they operate. Where damage to ecosystems has resulted from low-resource agriculture, it is because CEB was not maintained, and this can occur at low population densities as well as at high population densities. Nevertheless, the need to maintain CEB does impose limits on production possibilities. It is hardly surprising, therefore, that recent modeling exercises that have attempted to measure population supporting capacity in Africa at different hypothetical levels of agricultural technology, on the assumption that existing production systems can accommodate modern technology transfers, have produced results that can only be described as absurd.[6]

THE TENDENCY TO DIVERSIFY PRODUCTION
WITH INCREASING POPULATION DENSITY

In the English agricultural revolution, the Norfolk ("four-course") system of management with its new crops of turnips and clover and its elimination of fallow represented a diversification of production, even though the main sources of farm revenue (wheat and barley) remained the same as under the old system. The significance of the new husbandry, however, was that the new crops acted in the rotation to enhance the productivity, and thus the

profitability, of these cash crops in that, when fed to
livestock, they produced large amounts of manure which
served to increase grain yields.[7]

The desirability of diversifying outputs of low-
resource agriculture in Sub-Saharan Africa has long been
recognized by nutritionists, whose special knowledge makes
them concerned with what products of the farm stay on the
farm as well as with what products are sold in markets.
Economists, for their part, have come only slowly to
concede that diversity of output may have some advantages
for the farmer in the tropics. The feeling continues to
be that African agriculture, like agriculture in the
Western world, should be moving toward monocropping on the
Western pattern.

Diversification of output is a characteristic of
African low-resource agriculture. Farmers' investments in
small-scale infrastructure like bunding aim at objectives
like diversifying the crop mix grown on the farm rather
than raising crop yields or output per man-hour, as would
be farmers' likely objectives in commercialized agriculture. The rice farmers of the Sierra Leone scarp-foot
zone described by Richards purposely graded their bunded
fields so as to plant a range of rice varieties--some
quick-growing, some of medium duration, and some of long
duration--"both to spread the workload and to ensure a
harvest season lasting over the greater part of the
year."[8] The diversification of output permits factor
productivities to rise.

In other words, diversification of output goes hand
in hand with efficient intensification of production.
Although the increasing diversification of output with
intensification would appear go against intuition, such
historical evidence as exists in Africa suggests that
mixed cropping systems do indeed become more complex as
rural population densities increase. Guyer describes how
in the traditional groundnut fields cultivated by Beti
women the density of groundnut plants has increased over
time and also intercropping of varieties of cassava, a
staple food, has become common.[9] In the compound fields
of Lagemann's high-density Nigerian village there were
nine different field crops grown, compared with six in the
medium-density village.[10]

We are accustomed to associating specialization of
output with agricultural development. Yet, because of the
sustainability problem, there is little secure knowledge
of the effects of such specialization in African condi-

tions. When it appears spontaneously, as for instance with the increased incidence of monocropping of cassava in the Kwango-Kwilu, it may be evidence of inefficient intensification. The combination of crops and inputs leads to a vicious circle of declining soil fertility and specialization of output, as described by Fresco.

Cassava is a staple crop of the Kwango-Kwilu regions of Zaire, which lie astride the border between the SAT and TRF zones. Although accurate yield data are hard to come by in these regions, Fresco estimates that average cassava yields declined from 12 tons per hectare in 1958 to 5.5 tons per hectare in 1981. This decline in yields has been accompanied by an expansion in area cultivated to cassava.

Progressive reductions in fallow lengths have increased the burden of weeding on the women farmers of the regions. Burning provides a partial, and temporary, solution for weed problems and facilitates clearing after grass fallow, so that an increase in frequency of burning has been observed. To offset declining fertility, farmers have sometimes resorted to increasing field sizes, and, without exception, to planting cassava, which can be grown on depleted soils with low field-labor inputs. Since population densities are relatively low, land is readily available to farmers.

Techniques such as incorporation of plant residues have been abandoned because farmers consider them too labor-intensive to be applied to large surfaces. The incidence of diseases and pests has increased, and this may be related to environmental factors, such as shortened fallows, declining soil fertility, and cassava monocropping. These trends were undoubtedly aggravated by the system known as the "cultures imposées" by which certain minimum acreages for major crops have been enforced for the past generation (for tax collection purposes). This system means that fields are so large that the women who cultivate cassava, as they themselves say, "have no time to dig in grasses, [and] hoe and weed properly."[11]

Women resort to opening up new fields to take advantage of their higher initial soil fertility in preference to working old fields year after year. But without labor investment in CEB, y_2 declines and so do annual crop yields. On the horizontal axis of figure 5.4, the women farmers of the Kwango-Kwilu are actually moving to the left. The trends described by Fresco imply that this agricultural system, while presently supporting what is still a relatively sparse population (23 persons per square kilometer), may be expected to reach its critical population density in the very near future, if it has not already reached it. The "prosperity" of the Kwango-Kwilu regions is being propped up by sales of its surplus production of cassava to Kinshasa, which provides a ready market for it. But the effects of the land degradation that is taking place will be lasting on both the countryside and its people.

As Fresco says:

> One may speculate that if current trends continue, the final stage of shifting cultivation, at least on the Kalahari plateau, will be monocropping of cassava on exhausted soils.[12]

The Kwango-Kwilu case illustrates in a particularly sharp manner both the non-congruence of population density and resources that one often finds in Africa, and the tendency for monocropping to be equated with inefficient intensification of production. In the perspective of the development of low-resource agriculture that can accommodate rural population growth with productive employment of labor, therefore, we need to think in terms not only of resource conservation but also of increasing the diversity of output.

NOTES

1. As described in the writings of Pélissier and others. (Paul Pélissier, Les Paysans du Sénégal (Saint-Yrieix: Imprimerie Fabrègue, 1966).)

2. Boserup, reacting to the extreme Malthusian statement, put the case for causality between population growth and agricultural development most strongly. See Boserup, The Conditions of Agricultural Growth, especially pp. 11 and 117. In functional terms, the two statements may be expressed as follows:

Malthus: $Pg = f(Fs)$

Boserup: $Fs = g(Pg)$

where Pg = Population growth and Fs = Food supply.

3. Lauren A. Chester, "Cameroon: Population Growth and Land Resources, a Case Study," World Bank, mimeo, 1986.

4. Allan, The African Husbandman, p. 89. Italics in original.

5. Claude Raynaut, "Outils Agricoles de la Région de Maradi," chapter in ORSTOM, Les Instruments Aratoires, pp. 505-536.

6. See, for example, G. M. Higgins, A. H. Kassam, L. Naiken, G. Fischer, and M. M. Shah, Potential Population Supporting Capacities of Land in the Developing World (Rome: Food and Agriculture Organization (FAO), United Nations Development Program (UNDP), and IASSA, 1982). The authors calculated that Africa should be able to support 2.7 times its actual population at a low technology level, 10.8 times as much with an intermediate technology level, and 31.6 times as much with a high technology level.

7. Timmer, "The Turnip, the New Husbandry."

8. Richards, Indigenous Agricultural Revolution, p. 104.

9. Jane I Guyer, "Family and Farm in Southern Cameroon," African Research Studies No. 15, Boston University, African Studies Center, 1984, pp. 82-90. In the case of the Beti, what specialization occurred was in the increasing separation between men's crops and women's crops.

10. Lagemann, Traditional African Farming Systems, p. 35.

11. Fresco, Cassava in Shifting Cultivation, pp. 133-135 and 187-188; and Fresco, "Women and Cassava Production," mimeo (20 pages), 1982, passim. The author provides population density figures by local district in map 4, p. 76.

12. Fresco, Cassava in Shifting Cultivation, p. 210.

Conclusion and Implications

That mixed cropping systems of the African type are in most cases inimical to dramatic technological change has long been appreciated in disciplines other than economics. But economists have been slow to grasp the crucial importance to the sustainability of agricultural production in Africa of maintaining the soil biomass, and thus of fostering the development of mixed cropping in innovative ways. The mainstream view among economists that "conservation" models of agricultural development applied well to pre-19th century Europe but are outmoded today discouraged initiatives to break away from "technology-centered" models of agricultural development.

Yet the central obstacle to agricultural development in Sub-Saharan Africa has been the difficulty of bringing about intensification. In other words, how can resources be used more intensively in production? The problem manifests itself most immediately in the slow pace of technological change in African agriculture. But in reality it is a more general problem.

Rapidly diminishing marginal productivity of labor follows efforts to intensify agricultural production based on innovations that raise crop yields but leave all else unchanged, or that try to substitute monocropping for mixed cropping with the objective of raising yield of the principal crop alone.[1] This is because of the substantial increase in labor requirements, not only for cropping but also, more importantly, for conserving equilibrium biomass. There have been few instances in African agriculture where the yield of the principal crop was susceptible to being raised in sufficient degree to compensate for this expansion of labor requirement. As a

result, such efforts to intensify agricultural production
have been technically inefficient and efforts to get
farmers to adopt them have failed.

This book has shown how the intensification process
can work successfully in low-resource African agriculture
to raise productivity and to sustain production over the
long term. In the production function postulated here, on
the basis of documented evidence from farmers' fields,
conservation measures applied to the biomass determine the
sufficiency of the equilibrium attained for sustaining a
constant or rising trend of annual output. Such conservation measures depend heavily on the nature of the crop
mixture, as we have seen.

RETHINKING THE ROLE OF TECHNOLOGY

Where does this leave us with respect to the thesis
that low-resource agricultural systems can only be
improved through technological change, the question with
which this book began? Delgado and Mellor, in a recent
article, raise the provocative question of what constitutes the "adequacy" of a farming system's technology, since
they impute a low price elasticity of supply in agriculture in Sub-Saharan Africa in part to "the inadequacy of
existing technologies for lowland semiarid and humid
areas."[2] This seems to me to reflect a rather narrow If
self-confident) view, looking at agricultural technology
much as a builder of engines looks at cars. The
implication is that a more powerful technological "engine"
must be the way to get African agriculture moving.

Taking a conservation-oriented viewpoint in
consonance with the tried economic notion of efficiency,
and bearing in mind the important time dimension of
agricultural production (which, in fact, is the determinant of the outcome so far as African production trends
go), I would suggest that we need to readjust our thinking
about the role of technology. As originally conceived,
technology in agriculture was the whole bundle of tools
and biological materials that make production possible.
It is not necessarily so that technology is embodied in an
individual farm implement or crop variety. But if we
think it is, we are constantly perplexed by the failure of
African farmers to change their hand hoes or to plant this
or that new high-yielding variety recommended by extension
agents. And again, we are misled into believing that a

more powerful engine will make the car go, when what is needed is a tire that doesn't deflate at speeds over 25 miles per hour.

If we think of technology as organically linked to the whole system of growing crop mixtures, including the land clearing, fallowing, soil protection, and other management practices, which are inseparable from the classical inputs land, labor, and capital, we may think of the Green Revolution as having been a special case of innovation. In the Green Revolution, in particular, a very narrow subset of technological variables (high-yielding varieties of wheat and rice) were successfully applied to an environment covering a relatively broad geographic area (irrigable land in the tropics of Asia and Latin America) with dramatic effect.[3]

Two important implications follow: First, innovation is going on in African low-resource agriculture all the time as farmers adjust their cultivating practices and crop mixes to their soil and other resources, including labor. But it has been difficult to distinguish such slow, ecologically sound change from stagnation. Secondly, any technological innovations introduced from the outside must be evaluated, in the first instance, for their impact on resource conservation, rather than solely for their effects on crop yields or other productivity measures in the annual production process. More specifically, any changes introduced into African low-resource agricultural systems from the outside need to be evaluated for their effects on soil fertility and all the other diverse agronomic aspects of the resource base. Proper analysis of this phenomenon by economists was late in coming, perhaps due to the fact that the classical farm management economics approach focused on the relationship of inputs like land, labor, and tools to one another and to output, not on the effects these inputs have on the agricultural resource base, which is subsumed in the usual production function.[4]

In the theoretical analysis presented in this book, changing CEB acts as a multiplier of other factor productivities much in the manner of changing the technology parameter of the Hayami-Ruttan metaproduction function. If the analysis is correct, there is a strong likelihood that African low-resource agriculture is not producing at its peak capacity. Consequently, there is room for further improvement in factor productivities by facilitating the intensification process, even if bringing about

dramatic change is much more difficult than it was in the
Green Revolution.

Although the technological problem is quite different
in Africa, it is not at all beyond the application of the
principle of introduction and diffusion of innovations.
The technology to be applied is very broad, covering a
multitude of crops, cultivating practices, and tools, and
the environment is equally variable. Accelerating the
rate of intensification of agricultural production in
Sub-Saharan Africa implies finding the right matches
between the two.

The question is how.

THE RESEARCH AGENDA

While African low-resource agriculture may appear at
first sight to be tightly constrained, it possesses one
great advantage, as we have seen: its accommodation of a
wide range of cropping mixtures and sequential agricultural operations. The prevalence of mixed cropping
systems in African agriculture has been shown through the
evidence presented here, and its economic rationality for
sustaining annual output at a relatively stable level has
been demonstrated in terms of the economic analysis of
input use by means of a simple model of production
relations. The meta-isoquant construct, in particular,
offers insight into the flexibility which these cropping
mixtures afford in terms of alternative resource uses.

Research on African agriculture needs to be pursued
in the framework of the sustainability of agricultural
production. There have been very few studies by
agronomists, however, of the relationship between African
farmers' management practices and the sustainability of
agricultural production. Such studies require a long
period of observation. The study by Pichot and colleagues
from which the data in table 3.3 are derived is one of the
few I know of.

Sustainability studies have the potential to generate
sufficient data to permit economists to calculate partial
elasticities of output with respect to the factors of
production shared between output of CEB and annual crop
output, especially labor inputs. These elasticities can
then be compared with elasticities calculated from farm
survey data, and thus provide a basis for identifying
unexploited productivity potentials.

Diversification of production goes hand in hand with mixed cropping systems, and makes their operation easier. Although the increasing diversification of output with intensification of agriculture would appear to go against intuition, such historical evidence as exists in Africa confirms it.

In view of the rather recent arrival of mixed cropping on the agenda of agricultural research in Africa, the thinness of the body of systematic knowledge about it should not be too surprising. In the SAT of West Africa, although some research on mixed cropping dates to the 1930s, it was not until the 1960s with Norman's pioneering investigations of farming in northern Nigeria that mixed cropping began to be taken as a significant integral component of African agricultural systems. This was apparently due in part to the fact that the earlier work was not extensively reported.

Between 1930 and 1960 scientists at the Institute of Agricultural Research in Nigeria reportedly conducted over 300 experiments on crop mixtures. It was also in the mid-1960s that the ecological role of the acacia albida tree began to be scientifically investigated. Studies of mixed cropping in countries like Mali, Niger, and Senegal have been sporadic.[5] As late as 1971, Igbozurike's call for "unstinted research into mixed cropping" went virtually unheeded.[6] Most important, there has been an absence of continuous observation of any one cropping system to study how it has accommodated to a changing environment.

Once the desirability and need for accelerated research into mixed cropping have been accepted by the international agricultural research community, the general objective for improving such systems should be to maximize their flexibility so as to reinforce their sustainability. A number of fairly inexpensive ways of doing this have already been suggested by collaborating economists and agronomists.

Stewart and Kashasha, for example, working with rainfed agriculture in Kenya, have successfully translated rainfall availability patterns into feasible cropping combinations.[7] In this day and age, using modern communications in Africa to get up-to-the-minute meteorological information to large numbers of farmers should present no great difficulty, giving them greater flexibility of cropping decisions.

The land equivalent ratio (LER) is an important functional measure of the efficiency of low-resource agriculture. However, there have as yet been few studies in Africa of the relationship between the number of crops planted together and LER. Since it seems likely that one way of raising LER is by increasing the number of crops in the mixture,[8] one area of priority in research into ways of intensifying low-resource agriculture needs to be finding multiple cropping systems with as large a number of components as possible, including tree crops.

Increasing rates of fertilizer use offers a means of intensifying low-resource agricultural production in Africa, within the limitations of nutrient loss and other chemical effects on the soil described earlier in this book. However, applying fertilizers to crop mixtures is not the same thing as applying them to monocultures. As Steiner notes, the component crops have different nutritional needs and the period of maximum demand for one crop does not necessarily coincide with that of the associated crop(s). The application of nitrogen to a cereal/legume intercrop, for example, will decrease the use efficiency of nitrogen, as it suppresses symbiotic fixation by the legume crop. The same holds true for a maize/sweet potato intercrop, where nitrogen reduces the root yields of sweet potato. Application of phosphorus to a maize/cassava intercrop will result in a low use efficiency of phosphorus since cassava hardly responds to this nutrient.[9]

The problem is more complicated than this, however. Growth patterns of single crops change when they are grown in mixtures. For example, dry matter production by pigeonpea in a maize/pigeonpea intercrop was less than half that of sole cropped pigeon pea during the first 16 weeks. Once the maize matured, however, its competitive influence was reduced and the growth of the intercropped pigeonpea between the 16th and 24th weeks was sufficient to produce seed yields comparable to the sole crop. Shading is also a factor, like nutrient uptake, that depends on the nature of the crop mixture and its growth pattern. The evidence leads Steiner to state that no conclusion on the fertilization of cropping mixtures can be derived solely from information on the fertilizer requirements and response to certain nutrients of their components in sole stands.[10]

The improvement of African low-resource agriculture poses very stringent needs in terms of plant breeding. Probably in no other aspect of research are the differences in African requirements with those of the

Green Revolution biology more striking. Nevertheless, sorghum and millet varieties with promise for West Africa coming out of the ICRISAT program are given an early systematic screening only for date of planting response, not for their fit to land types or to intercropping.[11]

Cropping patterns are part of the agricultural environment. This implies that the problem of breeding and selecting new cultivars of African crops is complicated by the proliferation of environments in part because of the large number of cropping mixtures. The difficulty of fitting new cultivars into farmers' cropping patterns is a major one. As we have seen, it was finally an insurmountable obstacle to introduction of high-yielding cotton varieties into southern Nigeria in the 1930s, and it has been a similar obstacle to the success of many recent development projects in Africa. Yet there are many rational research programs to be suggested here. In the SAT zone of West Africa, for example, some national and regional research programs are selecting cowpeas for their suitability for intercropping with sorghum or maize. In the humid parts of Cameroon, where maize/cocoyam mixtures are common, maize is already being bred for its suitability to intercropping with this tuber crop in a small program.[12] Other modernized imitations of traditional practices, like tied ridging and alley cropping, have a bright future.[13]

In fallowing systems, since there is no unique relationship between CEB output and length of fallow period, research must be concentrated in the first instance on those areas where CEB can be maintained at least cost. These areas are not necessarily the same as areas of highest soil fertility, since in some cases maintaining the fertility of initially high-fertility soils can be extremely costly. As Richards points out, compared with sole-cropping systems the percentage gains from intercropping are greater on soils of lower rather than higher fertility.[14]

Commodity-centered research, such as has been carried on in the international agricultural research centers, focuses automatically on the areas endowed with the most favorable environment for growing the particular crop of interest, and especially areas where high soil fertility is likely to produce the largest payoff to new varieties of that crop. Other factors, like labor availabilities and productivity levels which depend on other components of the farming system, tend to be minimized or disregarded altogether. The assumption frequently made is that these matters are subordinate to the adoption decision (for reasons that appear logical to a Western viewpoint, such as the attractiveness of the higher income offered by the new crop variety), whereas in fact they are determinants of it.

In response to repeated failure of this sort, the tendency of the followers of the orthodox approach has been to enlarge the area where their priority crops may be adaptable to physical growing conditions (by sacrificing high yields for greater reliability, for instance), but not to confront the fact that their approach is an upside-down one from the point of view of the farmer.

What one often finds in Africa is that the experiment-station-package approach ends up being rejected by farmers for quite simple reasons that are obvious to farmers but may remain beyond the intuition of scientists working in experiment stations. In Sierra Leone, for instance, the labor demanded for work on water-controlled rice plots with a new technology package promoted by World Bank-financed Integrated Agricultural Development Projects conflicted with work on the household upland farm. Work on traditional early rice, private plots of floating rice and groundnuts, etc., and dry-season cultivation of swamps, on the other hand, tended to complement rather than compete with household farm work.

Where intensification does not seriously disrupt labor and other input use patterns, it takes place spontaneously, as it were. Many of the formerly uncultivated inland valleys and swamps in northern Côte d'Ivoire, for instance, have been developed in recent years by farmers as ricelands, some with sophisticated systems of leveling, bunding, irrigation, and drainage to facilitate good water control. Small reservoirs have even been built to provide year-round irrigation.

That the commodity-centered approach is wrong for low-resource agriculture should now be clear. The alternative research approach that is needed is based on what observers like Richards and Matlon suggest should be called "participatory research."[15] Learning the lessons of past experiences is a fairly simple matter and need not involve large budgets for elaborate research stations. But it does involve a particular mental aptitude on the part of researchers, who must be willing to learn from farmers. Their body of experience and knowledge is much wider than that of most researchers.

As Richards has written:

> In the community in which I am currently working in
> central Sierra Leone, the majority of farmers have,
> by middle age, made fifteen or twenty different farms
> giving them experience of three major soil types, two
> geological zones, and three vegetation systems. In
> addition, as children they became familiar with all
> the farms made by their parents, and, while members
> of labor cooperatives, gained first-hand experience
> of twenty or so farms a year in addition to their
> own. As shifting cultivators, therefore, they
> rapidly acquire detailed knowledge of practically
> every square meter of 37 km^2 of extremely varied
> terrain, and of the likely reaction of various soil
> formations and crop combinations to a range of cli-
> matic and hydrological contingencies. The experience
> of shifting cultivation has given these farmers a
> practical, comparative, knowledge of a wide range of
> ecological processes which might be the envy of the
> modern college graduate managing 1,000 hectares or so
> of highly uniform Kansas or East Anglian grain land
> let alone the European peasant of former times boxed
> into a tiny inherited plot or shackled to the land
> utilization rules of the feudal manor.[16]

THE DEVELOPMENT AGENDA

By a conservative estimate, there are 200 million
low-resource farmers in Sub-Saharan Africa. They
represent a potent force for growth in production and for
growth in consumption. Rural Africa constitutes a huge
market for agricultural products, and particularly food.
In Kenya, 80 percent of the total market demand for maize
comes from the rural sector.[17] The proportion is
probably similar for food crop demand in other African
countries.

Growth in productivity in low-resource agriculture
can be very broadly based due to the similarity of the
principles underlying African low-resource farming
systems. Such growth may not necessarily take the form of
quantum leaps in aggregate output or even in output of
specific crops. The cotton experience in francophone
African countries is undoubtedly the exception to the
rule. The growth pattern is more likely to be incremental
in nature.

The experience of a quarter century of agricultural development projects in Sub-Saharan Africa overwhelmingly shows that what has been based on imported tools, machines, seeds and other inputs, managerial skills, and capital invariably fails, while what has been based on locally made implements, inputs available with transport arrangements no more complicated than a donkey cart, farmers' skills, and capital saved on site invariably survives. On grounds of cost alone, development has to be affordable to African low-resource farmers, which means it has to be ultra-low in cost at the point of implementation. This implies a premium on programs based on self-help at the individual, household, or village level.

In the rural sectors of Sub-Saharan African countries, the key to development is income generation. We have defined low-resource agriculture as using no purchased inputs, so the terms of trade of African low-resource farmers cannot be calculated meaningfully in the usual sense. This definition must now be relaxed somewhat to allow the analysis to encompass the full range of economic forces. It is likely that as farmers progressively acquire income they will look further afield for command of productive resources. Furthermore, surplus output is marketed, and often at relatively high prices. This means that as the efficiency of production of low-resource agriculture increases, the incipient terms of trade between outputs and inputs exert a multiplier effect on the productivity ratio to increase returns over costs. This acts as an income generator in the rural sector.

At the same time, this intensification of production carries with it the possibility of greater labor use, which in the aggregate will absorb much more labor than large-scale commercial farming in the most fertile areas of Africa like the Nile irrigation schemes in Sudan. This will help stem rural-urban migration.

The burden on those concerned with development in Sub-Saharan Africa need not be so terribly great provided a correct approach to the problem is made. Such an approach needs to be based, obviously, on an accurate analysis of production relations. Increasing efficiencies in the use of resources in Sub-Saharan Africa's mixed cropping systems must then be coupled with appropriate pricing incentives and policies. These latter must be designed with the conservation of resources in mind. This will take time, because in-depth understanding of low-resource agricultural systems is scarce not only in foreign development agencies, but also in African governments because of the wide gap that has grown up between the rural and urban sectors of most African countries. Thus, a massive educational task confronts us in the years and decades ahead.

The worst outcome would be for the present impression of the "stagnation" of African agricultural methods to be allowed to stand unchallenged as the prevailing view. Low-resource agriculture in Sub-Saharan Africa is not stagnant, only highly resistant to imposed change. Among the professionals most qualified to change the "stagnation" view are the agronomists and soil and crop scientists of all kinds who work closely with the African physical environment. Economists, for their part, have an important role to play by focusing attention on the human element in these systems--farmers--and how they achieve efficiency of production and respond to economic variables, and above all prices. There is no need to be pessimistic, just because the efforts both of agronomists and economists so far have failed to make a major impact.

NOTES

1. A problem already identified by J. H. Cleave (African Farmers: Labor Use in the Development of Smallholder Agriculture (New York: Praeger, 1974)) and by J. Levi and M. Havinden (Economics of African Agriculture (Harlow: Longmans, 1982)).

2. They make the following statement:

A major reason for low supply elasticity in Sub-Saharan Africa is the poor state of agricultural infrastructure and input distribution systems in Africa . . . This is, of course, reinforced by difficult resource conditions, initially low input intensity, and the inadequacy of existing technologies for lowland semiarid and humid areas.

(Delgado and Mellor, "A Structural View," p. 667.)

3. Yujiro Hayami, "Conditions for the Diffusion of Agricultural Technology: An Asian Perspective," The Journal of Economic History, Vol. XXXIV, No. 1 (March 1974), pp. 131-148.

4. I am indebted to William I. Jones and Roberto Egli for this particular formulation of the problem.

5. This brief review is based on L. K. Fussell and P. G. Serafini, "Crop Associations in the Semi-Arid Tropics of West Africa: Research Strategies Past and Future," chapter in Ohm and Nagy (eds.), Appropriate Technologies, pp. 218-235.

6. Matthias U. Igbozurike, "Ecological Balance in Tropical Agriculture," The Geographical Review, 61 (1971), p. 529.

7. See J. Ian Stewart and D. A. R. Kashasha, "Rainfall Criteria to Enable Response Farming through Crop-Based Climate Analysis," contribution from the USDA/USAID/GOK Dryland Cropping Systems Research Project (615-0180), Agriculture Research Department, Kenya Agricultural Research Institute (ARD/KARI), Muguga, Kenya, November 1983. For examples of traditionally observed weather indicators in West Africa, see Richards, Indigenous Agricultural Revolution, p. 47.

8. See the evidence from Costa Rica in the form of figure 20 in Steiner, Intercropping, p. 85.

9. Ibid., p. 119.

10. Ibid., p. 130.

11. Matlon, "A Critical Review," pp. 158-159.

12. Steiner, Intercropping, pp. 104-105.

13. The tied-ridging system of the Dogon of Mali was described in a 1967 study by Henri Raulin, La Dynamique des Techniques Agraires en Afrique Tropicale du Nord (Paris: Etudes et Documents de l'Institut d'Ethnologie). The role of acacia albida in soil fertility maintenance in the West African SAT was described in Claude Charreau and P. Vidal, "Influence de l'Acacia Albida Del. sur le Sol, Nutrition Minérale et Rendements des Mils Pennisetum au Sénégal," L'Agronomie Tropicale, 20 (1965), pp. 600-626.

14. Richards, Indigenous Agricultural Revolution, p. 70.

15. In Matlon's words:

There is an urgent need for more emphasis on off-station research using well-placed researcher-managed trials and farmer-managed tests. Greater farmer participation is required at various stages in the development and testing of technologies to ensure greater farm-level adaptation. In particular, the principal factors causing yield shortfalls between the research station and farmers' fields need to be identified and fed back to modify as necessary on-station objectives and methods.

(Peter J. Matlon, "The Technical Potential for Increased Food Production in the West African Semi-Arid Tropics," paper presented at the Conference on Accelerating Agricultural Growth in Sub-Saharan Africa, Victoria Falls, Zimbabwe, August-September 1983.)

16. Paul Richards, "Ecological Change and the Politics of African Land Use," The African Studies Review, Vol. 26, No. 2 (June 1983), pp. 56-57.

17. Guenter Schmidt, "Maize and Beans in Kenya: The Interaction and Effectiveness of the Informal and Formal Marketing Systems," Occasional Paper No. 31 (Nairobi: Institute for Development Studies, University of Nairobi, 1979).

Index